BOND ORDERS AND ENERGY COMPONENTS

Extracting Chemical Information from Molecular Wave Functions

BOND ORDERS AND ENERGY COMPONENTS

Extracting Chemical Information from Molecular Wave Functions

István Mayer
Research Centre for Natural Sciences
Hungarian Academy of Sciences
Budapest, Hungary

CRC Press
Taylor & Francis Group
Boca Raton London New York

CRC Press is an imprint of the
Taylor & Francis Group, an **informa** business

CRC Press
Taylor & Francis Group
6000 Broken Sound Parkway NW, Suite 300
Boca Raton, FL 33487-2742

First issued in paperback 2021

ISBN 13: 978-0-367-86484-2 (pbk)
ISBN 13: 978-1-4200-9011-6 (hbk)
ISBN 13: 978-1-315-37489-5 (ebk)

DOI: 10.1201/9781315374895

Library of Congress Cataloging-in-Publication Data

Names: Mayer, István, 1943-
Title: Bond orders and energy components : extracting chemical information from molecular wave functions / István Mayer.
Description: Boca Raton : Taylor & Francis, 2017. | "A CRC title." | Includes bibliographical references and index.
Identifiers: LCCN 2016017525 | ISBN 9781420090116 (alk. paper)
Subjects: LCSH: Wave functions. | Chemical bonds. | Chemical structure. | Quantum chemistry.
Classification: LCC QC174.26.W3 M39 2017 | DDC 541/.22--dc23
LC record available at https://lccn.loc.gov/2016017525

Visit the Taylor & Francis Web site at
http://www.taylorandfrancis.com

and the CRC Press Web site at
http://www.crcpress.com

Publisher's Note
The publisher has gone to great lengths to ensure the quality of this reprint but points out that some imperfections in the original copies may be apparent.

Contents

Preface

Chemists consider a molecule as consisting of atoms held together by chemical bonds; physicists treat it as a system of individual particles—nuclei and electrons. From that perspective, most of quantum chemistry is basically a special chapter in computational molecular physics, rather than a branch of chemistry in the traditional sense of the word. Both approaches have their well-established fields and merits. If we need accurate numbers of bond lengths, formation energies, energy barriers, spectroscopic data, etc., then we have to calculate the wave functions (as accurately as we can) which means that we have to deal explicitly with the particles constituting the molecular systems. At the same time, all the enormous amount of chemical knowledge is systematized on the basis of the observation that there are similar molecules having similar atoms and bonds, which in many cases behave similarly both in the laboratory and in the living body. While nobody casts doubt that all these properties in the final count are determined by the behavior of the particles forming the molecule, we often cannot see that connection in any explicit manner. For instance, it would hardly be possible to explain what is, say, a secondary alcohol, referring directly to the electrons and nuclei constituting the molecule in question.

One would like to bridge this gap between the physical and chemical descriptions of molecular systems to make them constituents of a common thesaurus of the knowledge, instead of separate entities. Obviously, it would be best if one could *derive* the basic chemical laws and tendencies from the underlying physical theory (the Schrödinger equation). This seems, however, not realizable at this time, and there seems no real hope that this is to be changed soon. In such a situation one has to refer to the *a posteriori* analysis of the wave functions obtained in the actual calculations, and develop theoretical tools that permit connecting the results of calculations with genuine chemical concepts.

I started to work in this direction more than thirty years ago. My attention has been concentrated on the quantum mechanical meaning of the *bond order* (multiplicity of a chemical bond) and actual *valence* of an atom in a molecule, as well as the different atomic and diatomic *energy components* reflecting the energetics of formation of a molecule. Also, I paid great attention to the interesting problems of *effective atomic orbitals* within a molecule, usually forming an *effective minimal basis*, and more recently to the study of *local spins*.

The results that I obtained are scattered in numerous papers published in different journals, which makes it difficult to assess them, especially if their interrelations are concerned. The main aim of this book is, therefore, to put together these pieces in a systematic manner and to *present all the results with complete derivations.* For that reason, most of the results are obtained by using a common theoretical framework, the "atomic resolution of identity." In accord with my inclination, this book is basically about obtaining formulae, not their applications to actual chemical problems. Nonetheless, I hope that it may be useful for every quantum chemist interested in linking chemical concepts to quantum mechanics.

One of the basic difficulties in connecting the physical and chemical description of molecules is due to the lack of a unique definition of an *atom within a molecule.* As is discussed in the book, there are different definitions, none is perfect. Due to my preferences, most discussions are presented in terms of the so-called "Hilbert space analysis," when the atom is defined by its nucleus and basis orbitals. The great advantage of applying the "atomic resolution of identity" formalism is that it permits the formulae to be rewritten in a trivial manner also to the so-called 3D formalism, in which the physical space is partitioned into atomic domains. Accordingly, the 3D formalism receives less attention in the book.

Reading the book does not require any special knowledge, only materials every quantum chemist may be assumed to be familiar with. An exception is some sporadic use of a special second quantization formalism (use of creation and annihilation operators) for non-orthogonal orbitals. It is explained in detail in the appendix (along with the conventional second quantization for an orthonormal basis).

I use this opportunity to express my sincere gratitude to my coworkers—coauthors of several works discussed in the book—in particular to Drs. Pedro Salvador, Andrea Hamza, and Eduard Matito. Last but not least this book could not have been written without the permanent and crucial backing I receive from my wife, Dr. Márta Révész.

1

Introduction

1.1 Why a posteriori analysis?

The aim of performing quantum chemical calculations may be twofold. There are cases when we are just interested in getting a *number,* for instance to determine what conformer of a molecule is more stable or to estimate the barrier of a chemical reaction. Another purpose may be understanding or interpreting one or another property of the system studied—may be discussing the question of why the number just mentioned is as large or small as it is, and how it can depend on a substitution or other changes in the system. It is true that, in principle, any possible information concerning a molecular system is included in its *wave function*, but any *direct interpretation* of the latter is out of range of our capabilities in most cases of practical interest. This is the case because the wave function of a molecule is a multidimensional—may be complex-valued—function depending on a large number of variables which is very difficult to visualize even in the simplest cases. In the practical (necessarily approximate but more and more accurate) calculations the wave functions are represented by enormously big sets of numbers of different types—it is again hard to do anything directly with them. A further disturbing factor, discouraging the direct analysis of wave functions, may be connected with the fact that often *the same* wave function can be given in several, quite different but fully equivalent, forms; for instance it can be written down either by using the delocalized "canonical" molecular orbitals or in terms of some localized ones which can be assigned to different fragments of the molecule.

Another conceptual problem with the practical handling of wave functions is connected with the fact that it is very difficult to compare the results for different—even chemically very similar—molecules: their wave functions are defined in Hilbert spaces of different dimensions. This is, too, connected with the basic problem of relations between quantum chemistry and "ordinary" chemistry: the wave function is pertinent to the individual particles—

electrons and nuclei—forming the molecular system and describes their *physical* behavior, but does not contain any direct reference to the fundamental *chemical* notion of treating molecules as consisting of atoms kept together by chemical bonds. This chemical picture is both quite visual and well established, but bridging it with the wave function accounting for the physical behavior of the individual particles is far from trivial—mainly because there is no strict quantum mechanical definition of an atom *within* a molecule.

As the basic entity in chemistry is an atom, and there is no unique definition for the concept of an atom within a molecule, we have to introduce one in order to analyze the wave function, which aims to make links to the genuine chemical concepts. Of course, this definition inevitably contains an element of arbitrariness—a drawback which can hardly be avoided at present. There are two main possibilities for defining the atom, leading to two main avenues of the analysis [1]. In one case, we shall call it "Hilbert space analysis," one utilizes the fact that most quantum chemical calculations are performed by using atom-centered one-electron basis sets, and identifies the atom with the nucleus and the subspace of the basis functions assigned to that atom. Another possibility is to decompose the real three-dimensional (3D) physical space into atomic regions ("domains"). The atomic domains can be either disjunct (non-overlapping)—the most important case of which is Bader's [2] "atoms in molecules" (AIM) theory*—or "fuzzy" (lacking sharp boundaries) as has been introduced by Hirshfeld [3].

Accordingly, it is of both practical and conceptual importance to subject the immense amount of numbers produced in quantum chemical calculations to a sort of special "data compression"—such that it permits one to obtain some *chemically* meaningful quantities that may be put into correspondence with the respective classical chemical concepts. Such parameters calculated *a posteriori* from the results of quantum chemical calculations should give some important conceptual links between the physical and chemical ways of describing nature and can be very useful for interpretation and systematization of the results of calculations and understanding the properties of the systems studied. Also, if treated properly, they can have some predictive power, too.

In our opinion, the most important quantities of the above type are the electron population (or resulting charge) of an atom, the bond order (multiplicity)

*Bader's atomic domains are defined by the topology of the electron density and have a number of special properties making them attractive from a theoretical point of view. However, we cannot accept Bader's claim that they represent the *only* correct definition of an atom within the molecule. An important objection is that Bader's analysis often results in additional domains containing no nuclei (so called "non-nuclear attractors") which cannot be identified with any of the atoms and are, therefore, lacking any chemical significance.

between individual pairs of atoms and the components of the total energy which one can assign to the individual atoms and pairs of atoms. This book deals mainly with these three types of quantities and represents an attempt to systematize the author's results in that field, obtained in the last three decades. Some promising results have also been obtained for an interesting quantity, the *local spin*, simultaneously with preparing this book; they are also included and systematized here for the first time.

1.2 A prelude: The simple Hückel theory

Quantum chemistry as a field representing real interest for practical chemists essentially began with the simple (but ingenious) Hückel theory, and theoretical organic chemistry is to a large extent based on the notions introduced in that framework. But a number of concepts applied in the Hückel theory (e.g., those of HOMO, LUMO) have a much wider use, too.

As is known, Hückel theory deals only with the π-electron system of conjugated planar molecules, by assigning one π-orbital to each atom which is treated as a "π-center," while the σ skeleton of the molecule is not explicitly considered. The incredible simplicity of the Hückel method helped it to win the "competition" with the classical VB ("valence bond") theory which took place in the early times, although the VB scheme operated with much more "chemical" concepts.* Besides simplicity, the key factor in the success of the Hückel method was the fact that in its framework Coulson solved three problems at once for the simple π-electron theory: determining the distribution of the electronic charge between the atoms, characterizing bond multiplicities, and presenting the π-electron energy as a sum of atomic and diatomic contributions. (The connection of Coulson's approach with the energy partitioning is less widely recognized although it was already discussed in his classical paper [4].)

The "translation" of the results of the Hückel calculations to a form intelligible for a human was performed by Coulson's "charge–bond order matrix." This nomenclature is not really appropriate beyond the π-electron theories, so in the general case one usually calls it "density matrix" or simply "P-matrix" or "D-matrix." (We shall use letter P for the spin-dependent and letter D

*However, we are obliged to the VB method for such basic concepts as the resonance between different hypothetical "ideal" (e.g., Kekulé) structures.

for the spin-independent quantities.) It is important to stress that this matrix is not an *ad hoc* auxiliary tool used for interpretations but it is the basic mathematical invariant determining (up to an unimportant constant factor) the single determinant many-electron wave functions, including those used in the Hückel theory.

In the Hückel method one obtains the molecular orbitals (MOs) by solving the eigenvalue problem of an effective one-electron Hamiltonian matrix (Hückel matrix, **H**). In the case of conjugated hydrocarbons this matrix **H** is nothing more than the adjacency matrix giving the topology of the π-centers; in the case of π-systems containing heteroatoms the situation is somewhat more complicated, because one has to introduce some parameters, but that is of no importance from our present point of view.

The Hückel eigenvalue problem, written down in components, is

$$\sum_{\nu=1}^{N_\pi} H_{\mu\nu} c_\nu^i = \varepsilon_i c_\mu^i \, . \tag{1.1}$$

where N_π is the number of basis orbitals (π-centers), c_μ^i is the μ-th component of the i-th eigenvector $\mathbf{c}^i$, and ε_i is the respective eigenvalue. This equation is also equivalent with the matrix diagonalization problem $\mathbf{C}^\dagger \mathbf{H} \mathbf{C} = diag\{\varepsilon_i\}$, assuming that the i-th column of matrix **C** is the eigenvector $\mathbf{c}^i$. (The dagger indicates the adjoint.)

The molecular orbitals φ_i are determined by the components of the eigenvectors:

$$\varphi_i(\vec{r}) = \sum_{\mu=1}^{N_\pi} c_\mu^i \chi_\mu(\vec{r}) \, , \tag{1.2}$$

where the χ_μ-s are the π-type atomic orbitals centered on the individual centers. (One usually considers them as $2p_z$ orbitals, assuming the molecule is oriented along the xy-plane.)

Although the orbitals φ_i and the eigenvalues ε_i represent the complete solution of the Hückel problem, it is very difficult to find out anything about the actual *chemical* structure of the molecule directly on the basis of the orbitals. For instance, by looking only at the orbitals one can hardly tell what bonds of a conjugated system exhibit a larger and what ones a smaller double bond character. These properties usually cannot be attributed to some individual orbitals but are determined by more subtle simultaneous changes in the coefficients of several orbitals. At the same time all such questions are immediately answered by Coulson's D-matrix, the elements of which are determined by the coefficients of all the occupied ("occ.") orbitals *collectively*.

Probably the simplest way of introducing the D-matrix in a systematic manner is based on the expansion of electron density. (Coulson's original definition was made more or less heuristically.) It is known—see e.g., page 228 in [5]—that for a single determinant wave function built up of doubly filled orbitals φ_i (we shall consider only such ones in this section) the spatial distribution of the electron density is given by a simple sum of the orbital densities:

$$\varrho(\vec{r}) = 2\sum_{i}^{occ.} |\varphi_i(\vec{r})|^2 = 2\sum_{i}^{occ.} \varphi_i^*(\vec{r})\varphi_i(\vec{r}) , \tag{1.3}$$

where, for the sake of generality, the case of complex orbitals is also admitted.* Substituting the expansion (1.2) of the MOs, one trivially gets

$$\varrho(\vec{r}) = \sum_{\mu,\nu=1}^{N_\pi} D_{\mu\nu}\chi_\nu^*(\vec{r})\chi_\mu(\vec{r}) , \tag{1.4}$$

where

$$D_{\mu\nu} = 2\sum_{i}^{occ.} c_\mu^i c_\nu^{i*} . \tag{1.5}$$

In matrix notations one has

$$\mathbf{D} = 2\sum_{i}^{occ.} \mathbf{c}^i \mathbf{c}^{i\dagger} , \tag{1.6}$$

a sum of dyadic products. We may note here that the form of Eq. (1.4) is valid for any wave function expressed in a finite basis, only the relationship $\mathbf{D}^2 = 2\mathbf{D}$ characteristic for single determinant wave functions built up of doubly filled orbitals expanded in an orthonormalized basis is not valid in the general case.

Integrating both sides of Eq. (1.4) over the whole space, by taking into account that the orbitals in the simple Hückel method are assumed orthonormalized, one gets for the number of electrons N

$$N = \sum_{\mu}^{N_\pi} D_{\mu\mu} , \tag{1.7}$$

indicating that the diagonal elements $D_{\mu\mu}$ can be identified with the electron population of the μ-th π-center.

*It is the author's opinion that when doing derivations it is worthwhile to admit the possibility of using complex wave functions. That helps one to keep track of what terms are coming from where and avoid confusion. However, when the programming work is considered, one has to take the advantage that the functions used in practice are usually real.

Coulson identified the off-diagonal elements $D_{\mu\nu}$ of matrix $\mathbf{D}$ with the π-electron bond order between the μ-th and ν-th π-centers. Assuming that all the σ-bonds are single ones, then one gets for the multiplicity of the given bond the value $1 + D_{\mu\nu}$. The legitimacy of this assumption may be illustrated by a few numerical examples. In the ethylene molecule the π-electron bond order of the C=C bond is exactly 1, so we have the case of a pure double bond; in butadiene the π-electron bond order of the formal double bonds has the value of 0.894 which is only slightly smaller than unity, while the π-electron bond order of the formal single bond—which is, however, conjugated with the double ones—is as large as 0.447, in full accord with the fact that this bond is significantly shorter than ordinary C–C single bonds. It is also in full accord with the chemical expectations that the π-electron bond order of the benzene is 0.667, or that for the infinite graphite plane one gets the value of 0.525 [6,7].

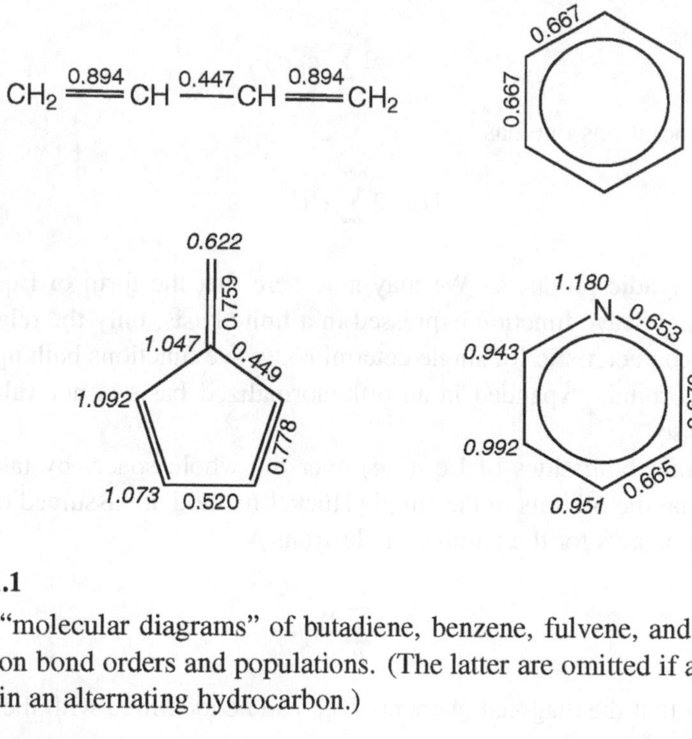

Figure 1.1
Hückel "molecular diagrams" of butadiene, benzene, fulvene, and pyridine: π-electron bond orders and populations. (The latter are omitted if all exactly equal 1 in an alternating hydrocarbon.)

These results permit to plot very informative "molecular diagrams"—like those on Figure 1.1 in which the π-electron populations (or net π-electron atomic charges) and the nearest-neighbor bond orders are indicated, permitting to present the basic results of calculations in a very compact and visual manner.*

The molecular diagrams were a widely accepted form of presenting the results for some time, and not only for the Hückel level of theory; even their compilations were published (e.g., [8]). Later the molecular diagrams gradually disappeared. That might be connected both with the absence of an appropriate measure of bond multiplicity in the *ab initio* theory and with the fact that optimization of molecular geometries became possible, and presentation of geometric data became prevailing. While it is commonplace that there is a correlation between the lengths of individual chemical bonds and their strengths, we do not think that bond length data can replace the chemical concepts of bond multiplicity and bond energy.

In the Hückel theory one does not consider explicitly the interaction between the electrons, so the total π-electron energy is simply the sum of the orbital energies of occupied orbitals, i.e., for the closed-shell case that we are considering

$$E_\pi = 2\sum_{i}^{occ.} \varepsilon_i \, . \tag{1.8}$$

The intimate connection between the "charge–bond order matrix" **D** and the components of the π-electron energy can be obtained by a simple derivation. We multiply both sides of the eigenvalue equation (1.1) by c_μ^{i*} and sum over i and μ; then we recognize on the left-hand side the definition (1.5) of the D-matrix elements, while on the right hand side we utilize the normalization of the MOs to get the total π-electron energy (1.8). Thus we get the known relationship $E = Tr(\mathbf{HD})$ expanded in components:

$$E_\pi = \sum_{\mu,\nu=1}^{N_\pi} H_{\mu\nu} D_{\nu\mu} \, . \tag{1.9}$$

This means that the energy is a sum of terms which may be systematized according to the indices μ and ν involved. Thus one can assign to the μ-th π-center the energy component $E_\mu = H_{\mu\mu}P_{\mu\mu}$, which is proportional to the π-electron population of that center, in full accord with the fact that no explicit electron–electron interactions are taken into account in the formalism. As the summations on the right-hand side of Eq. (1.9) run independently, there is an energy contribution corresponding to each pair of indices μ and ν:

$$E_{\mu\nu} = H_{\mu\nu}D_{\nu\mu} + H_{\nu\mu}D_{\mu\nu} = 2Re\{H_{\mu\nu}D_{\nu\mu}\} \, , \tag{1.10}$$

which is proportional to the respective π-electron bond order.

*Note that for "alternating" hydrocarbons—i.e., those lacking odd-membered rings—all the carbons are strictly neutral in the π-electron theories, so there is no need to indicate the charges for them.

These considerations indicate that the total π-electron energy of the Hückel method decomposes in the most natural manner into a sum of one- and two-center contributions:

$$E_\pi = \sum_{\mu=1}^{N_\pi} E_\mu + \sum_{\mu<\nu}^{N_\pi} E_{\mu\nu} \,. \tag{1.11}$$

These, however, are not often discussed explicitly as the individual energy components are simply proportional to the respective D-matrix elements. (This is not the case beyond the Hückel theory.)

Finally one should mention that the rather good correlation observed for the bond lengths between the sp^2 carbon atoms and the Coulson bond orders is obtained by using *the same* parameter values for each bond, i.e., without introducing the actual bond lengths in the calculations in any manner. This means that the Hückel method has an unexpectedly high *predictive power*, especially in light of its ultimate simplicity: it tells us that these bonds are short or long *because* they are strong or weak, and not simply they are found strong (weak) because they are short (long).

1.2.1 An application: Hückel's "4n+2 rule"

It is well known that cyclic conjugated hydrocarbons are aromatic if the number of π-electrons is equal to $4n+2$ ($n = 1,2,\ldots$) and antiaromatic if it is equal to $4n$. The enhanced stability of most different cyclic systems with 6 (10) active electrons is a quite general observation for both stable molecules and transition states, so it must have some general (topological) reasons. Therefore, simple Hückel level of theory, reflecting just the topology of the system, should be able to explain this phenomenon qualitatively.*

One commonly uses a very simple argument which is based on our general *belief* that electronic configurations with closed shells are particularly stable. In fact, the electronic levels of a regular Hückel polygon are distributed in the manner that after the fully symmetric lowest level follow pairs of doubly degenerate levels corresponding to two-dimensional irreducible representations of the symmetry group. The wave function of the molecule has the proper symmetry if the basis functions of each irreducible representation are either completely occupied or are left completely empty. Thus we have to fill an odd number of levels; as two electrons of opposite spin can be put on each level, this requires twice an odd number, i.e., $4n+2$ of electrons. However, the stability of closed shells, although fulfilled in most cases, cannot be *derived*

*Part of this section is reprinted from my paper [10] with permission from Springer.

in a truly general form, thus referring to it may be considered insufficient for a true explanation of the "$4n+2$ rule."

One can approach the problem from a different perspective [9, 10] which stresses the importance of the density matrix elements and of the respective (proportional to them in the Hückel case) energy components discussed above.

One of the most important aspects of aromaticity is that there is an energetically favorable *cyclic* delocalization, which is strong enough to prevent addition reactions, while for antiaromatic systems the cyclic delocalization is destructive (destabilizing). To judge about the effect of cyclic delocalization, one has to compare the open-ended and the closed Hückel chains of the same length. Thus one can start considering the problem by using the known solutions for the *open-ended* chains.

Coulson had solved analytically the problem of linear Hückel chain (see e.g., [11]). For an *open-ended* linear chain of the length $m = 2k$, the MO coefficients are given by his formula

$$C_{jr} = \sqrt{\frac{2}{2k+1}} \sin\left(\frac{jr\pi}{2k+1}\right), \tag{1.12}$$

where j is the index of the carbon atom and r of the MO (or *vice versa*—in this case matrix $\mathbf{C}$ is both orthogonal and symmetric).

Now we *close* the ring by introducing the respective $H_{1m} = H_{m1} = 1$ matrix elements (in $\beta < 0$ units). According to the relationship (1.10), the first order energy change will be [9]

$$\Delta E^{(1)} = 2D_{1m}. \tag{1.13}$$

That means that the Coulson's bond order *between the ends of an open chain* predicts the first order effect on the ring closure [9]. It is stabilizing if $D_{1m} > 0$ and destabilizing if $D_{1m} < 0$.

Using the coefficients in Eq. (1.12), one could calculate this bond order analytically [10,12], and it shows a characteristic sign alternation and leads to the "$4n+2$ rule" in all cases. In particular, for a neutral polyene of length $m = 2k$ one gets [12]:

$$D_{1(2k)} = \frac{(-1)^{(k-1)}}{2k+1}\left[1 + \sec\frac{\pi}{2k+1}\right]. \tag{1.14}$$

The angle in the argument of the secant ($\equiv 1/\cos$) is less than $\pi/2$, so the expression in the braces is positive. Therefore, Eq. (1.14) predicts that the ring closure is stabilizing if k is an odd number, and destabilizing if k is even.

The number of π-electrons is $N = m = 2k$, i.e., we get stabilization (possible aromaticity) if the number of electrons is twice an odd number $(4n+2)$, and destabilization (antiaromaticity) if it is twice an even number $(4n)$. This alternating behavior of the bond order $D_{1(2k)}$ permitted to classify the $4n+2$ polyenes as Hückel-type systems and the $4n$ ones as Möbius systems [12]. In an analogous manner one could *derive analytically* the "$4n+2$ rule" also for the double ions of even-membered chains as well as for the cations and anions of the odd-membered chains [10], predicting, for instance, the strong aromaticity of cyclopentadienyl cation.

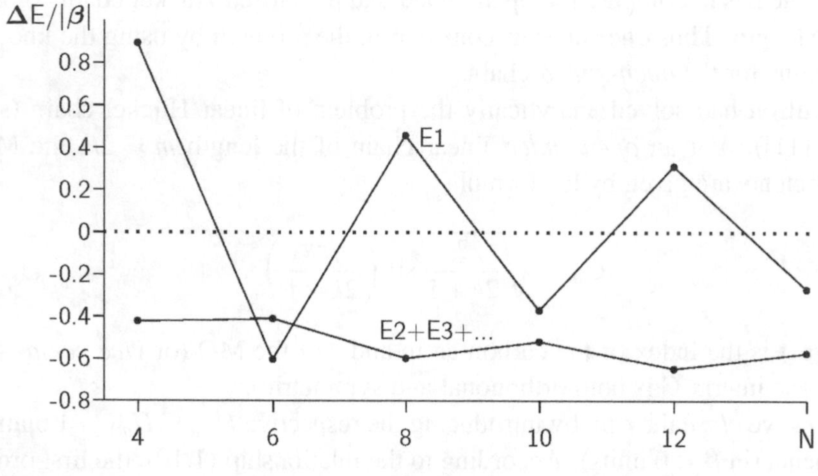

Figure 1.2
The first order and higher order energy contributions taking place when a Hückel chain is closed. [10] with permission from Springer.

Figure 1.2 shows that the contributions of the second and higher order in perturbation theory exhibit only a moderate alternation, so the difference in the behavior of the aromatic and antiaromatic systems is indeed determined by the sign and value of the ring-closing density matrix element D_{2m}: for systems not fulfilling the Hückel "$4n+2$ rule," the negative higher order energy contributions are not sufficient to catch up to the significant positive energy component observed at the first order, and the total energy change of ring closure is either positive (cyclobutadiene) or is not sufficient to compensate for the fact that the number of σ bonds decreases by one when closing the ring.

These strong alternations have a simple topological background, connected with the distribution of the nodes of the MOs. As is known, the orbital having the lowest energy has no nodes, the next has one node, the third two, and so on. Thus for the first level the coefficients at the two ends of the chain are of the same sign, giving a positive contribution to the density matrix elements D_{1m}; for the second level these coefficients are of opposite sign and give a negative contribution; the third again gives a positive one etc. Thus in butadiene there are two contributions of opposite sign. Simple considerations show that the negative one must have a larger absolute value. In fact, the conjugation (delocalization) in an open-ended butadiene chain leads to an energy lowering if and only if $D_{23} > 0$. This means that for the nodeless lowest state the coefficients at the positions 2 and 3 should be greater in absolute value than those of the second level for which there is a sign change. As the orbitals are normalized, the situation for the outer coefficients must be the opposite, leading to a negative D_{14} matrix element, and destructive cyclic delocalization in the cyclobutadiene. (Therefore it is unstable and contains two practically isolated double bonds.) In hexatriene there are two positive and only one negative contributions to D_{16}, and the former prevail. With the further increase of the chain we have a further alternating effect, even if with a decreasing amplitude. These qualitative considerations show that the results obtained above at the Hückel level must remain basically true for more refined schemes as well, and should reflect the actual chemical behavior of the systems in question.

2

Basic ideas of Hilbert space analysis

As noted in the introduction, one of the main avenues of analyzing the results of quantum chemical calculations makes use of the fact that most of the practical calculations apply atom-centered basis functions. This permits to define the atom within a molecule as a nucleus and the set of the basis orbitals centered on that atom.* This approach can be traced back to the early days of quantum chemistry, when the interaction between the individual atoms has been described by using the atomic orbitals of the free atoms to build up an approximate wave function. Hence comes the term "linear combination of atomic orbitals" (LCAO), which is also often used simply to denote the situations in which finite basis sets are used. A significant part of our "chemical thinking" is based on this approach—it is sufficient to recall here the distinction between *sp*, sp^2, and sp^3 carbon atoms; this characterization of the hybridization state of a minimal atomic basis tells us much about the atoms in question. Almost all of our qualitative considerations of the behavior (i.e., electronic state) of an atom in a molecule is based on the concept of a minimal *atomic* basis.

In the practical calculations one often loses direct contact with the qualitative LCAO picture, even if a minimal basis† calculation is used. For instance, it is hard to recognize the sp^3 hybrids in the output of an *ab initio* calculation of methane—even if the minimal STO-*N*G basis set is used. (The sp^3 hybrids cannot be strictly identified even if one turns to localized orbitals.) This is the case because, besides the main qualitative effects accounted for in the hybridization model, there are numerous smaller (but quantitatively not negligible) ones, which are also taken into account in an actual calculation, and it is hard to take them apart in the results.

In almost all practical calculations one uses (much) larger basis sets than the minimal ones, and for them it is utterly hopeless to look at the results

*This definition is easily extended to whole molecules when considering the problem of intermolecular interactions.

†The minimal basis sets applied in practice are not the free atoms' orbitals but their approximations with nodeless exponential functions (Slater-type orbitals, STOs) or rather the approximation of the latter with a combination of *N* Gaussians, (e.g., the STO-3G basis).

with the "naked eye" in order to find out an interpretation of the results in simple terms based on either the LCAO scheme or on any other chemical concept. However, the practice of using atom-centered basis sets permits one to develop different, more elaborate computational means of analyzing the numerical results. They are the means of the "Hilbert space analysis."

Of course, not all basis sets are equally well suited to perform an analysis of the results in chemical terms. The more "atomic character" a basis set has, the better it is for the analysis—in full accord with the classical LCAO notions. It is not sufficient that a basis set give a low energy. It should also bear some conformity with the chemists' atomic concept. Very "chemical" basis sets are e.g., the minimal STO-*N*G bases or the classical 6-31G** basis. (At the same time, the "split valence" *N*-31G type basis sets are less balanced.) There are adequate larger basis sets, too, e.g., the cc-pVTZ basis. It may happen that the use of too large basis sets makes it difficult to extract information of chemical type from the wave function or, at least, requires special measures to be taken. For instance, basis sets containing diffuse functions—i.e., basis function with rather small Gaussian (or Slater) exponents—that vary very slowly in the space and thus are lacking a well-defined atomic character, may lead to embarrassing results [13]. In some cases the solution may be to use a large basis set for energetic purposes (e.g., geometry optimization) and a smaller one for analyzing the qualitative features of the system. Lendvay suggested this approach especially with the aim to follow the progress of a chemical reaction: how one bond is breaking and another is formed when moving from reactants to the transition state and then to the products along the reaction path (e.g., [14–17]).

As to the diffuse functions, one can conclude that they are often important only from a quantitative point of view; there are, however, cases when the use of diffuse functions is essential for understanding the main features of the system, e.g., negative ions or the description of intermolecular interactions. In such cases the analysis may become much more difficult; turning to a Löwdin-orthogonalized basis and performing the analysis in it may often be the remedy.

It is worth mentioning here that the Hilbert space analysis also meets a conceptual difficulty which becomes of practical importance, too, if very large basis sets are used: the quantities it gives are lacking any complete basis limit. In fact, let us assume that one has a strictly complete basis on one atom. In that case any wave function can be expanded in terms of this "atomic" basis and there is no need for any basis functions on the other atoms. But then any quantity will be assigned to that atom. If moving the center of the basis to another atom, it gets that privileged role. But a mathematically complete basis may be centered anywhere, even on the Moon. This problem is an obvious manifestation of the fact that the concept of an atom *within* a molecule cannot be defined in any unique manner.

It is to be pointed out that the individual basis orbitals are of limited importance in some sense: any result of calculations must remain invariant if one replaces the initial set of atomic orbitals (AOs) centered on the given atom by their arbitrary linear combinations. This is the "rotational–hybridizational invariance" requirement,* the importance of which was probably first stressed by Pople when developing the CNDO (complete neglect of differential overlap) semiempirical scheme [18]. This means that only the *subspace* of the basis orbitals centered on the given atom has a true importance.

The subspace spanned by the basis orbitals $\{\chi_\mu\}$ attributed to a given atom A is uniquely defined by the *projector* (projection operator) $\hat{P}_A$ on that subspace. Because, in general, the basis orbitals are not orthogonal to each other even if they belong to the same atom, that projector can be written down by using Dirac's "bra" and "ket" notation as

$$\hat{P}_A = \sum_{\mu,\nu\in A} |\chi_\mu\rangle S^{-1}_{(A)\mu\nu} \langle\chi_\nu| \, , \tag{2.1}$$

where the symbol "$\in A$" indicates that the summation is extended only to the orbitals belonging to atom A, and $S^{-1}_{(A)\mu\nu}$ is shorthand for the element of the inverse atomic overlap matrix: $S^{-1}_{(A)\mu\nu} \equiv (\mathbf{S}^{-1}_{(A)})_{\mu\nu}$. (Similar shorthands will be used everywhere.) The form (2.1) of the projector we shall use throughout; it can easily be obtained by introducing an auxiliary Löwdin-orthogonalized basis $|\psi_\rho\rangle = \sum_{\mu\in A} S^{-1/2}_{\mu\rho} |\chi_\mu\rangle$ for the atomic subspace considered, and substituting this expansion and its adjoint into the more familiar form of the projector $\hat{P}_A = \sum_{\rho\in A} |\psi_\rho\rangle\langle\psi_\rho|$, valid in the case of orthonormalized orbitals. (The two matrices $\mathbf{S}^{-1/2}_{(A)}$ will combine into a single $\mathbf{S}^{-1}_{(A)}$.)

It is easy to check that the operator $\hat{P}_A$ defined by Eq. (2.1) satisfies the requirements of Hermiticity and idempotency

$$\hat{P}_A^\dagger = \hat{P}_A \, ; \qquad \hat{P}_A^2 = \hat{P}_A \, , \tag{2.2}$$

i.e., it is indeed a projection operator. Any linear combinations of the orbitals centered on atom A are its eigenvectors with eigenvalue equal unity,

*The term "rotational" refers to the fact that the rotation of the whole molecule with respect to the laboratory system of coordinates also induces a linear transformation between the basis orbitals of *p, d, etc.* type, centered on the same atom.

while any orbitals in the orthogonal complement to the subspace of atom A (i.e., orthogonal to all $|\chi_\mu\rangle$, $\mu \in A$) are annihilated, i.e., are eigenvectors with zero eigenvalue. In turn, the functions orthogonal to all basis orbitals of A are eigenvectors with eigenvalue unity of the operator $1-\hat{P}_A$ representing the projector on the orthogonal complement to the atomic subspace; while $(1-\hat{P}_A)|\chi_\mu\rangle = 0$ if $\mu \in A$. Obviously, the sum of these two projection operators is a unit operator, i.e., they together represent a "resolution of identity."

If one considers the molecular orbital

$$\varphi_i = \sum_\mu c^i_\mu \chi_\mu \,, \tag{2.3}$$

extending over the whole molecule, then it is natural to ask how large is the projection of it on the subspace of a given atom A. That question can be answered by calculating the square of the norm of the projection $\hat{P}_A\varphi_i$, i.e., the overlap integral of $\hat{P}_A\varphi_i$ with itself:

$$\|\hat{P}_A\varphi_i\|^2 = \langle \hat{P}_A\varphi_i|\hat{P}_A\varphi_i\rangle = \langle\varphi_i|\hat{P}^2_A|\varphi_i\rangle = \langle\varphi_i|\hat{P}_A|\varphi_i\rangle \,, \tag{2.4}$$

where the properties (2.2) have been utilized. Substituting here Eqs. (2.1) and (2.3), one obtains by a simple calculation

$$\|\hat{P}_A\varphi_i\|^2 = \sum_{\rho,\tau} \sum_{\mu,\nu\in A} P^i_{\tau\rho} S_{\rho\mu} S^{-1}_{(A)\mu\nu} S_{\nu\tau} \tag{2.5}$$

where $\mathbf{P}^i = \mathbf{c}^i\mathbf{c}^{i\dagger}$ is the "density matrix" corresponding to orbital φ_i; it has the elements $P^i_{\tau\rho} = c^i_\tau c^{i*}_\rho$.

Assuming the orbital φ_i normalized to unity, the quantity $\|\hat{P}_A\varphi_i\|^2$ gives the relative weight of its projection on the atomic subspace. As already noted in Section 1.2, the charges corresponding to the individual orbitals are additive for single determinant wave functions, so it is tempting to define—as has been done by Roby [19]—the quantity

$$Q^{\text{Roby}}_A = \sum_{\rho,\tau} \sum_{\mu,\nu\in A} D_{\tau\rho} S_{\rho\mu} S^{-1}_{(A)\mu\nu} S_{\nu\tau} \tag{2.6}$$

as the electron population of the atom A. Here $D_{\tau\rho}$ is an element of the total spin-less "density matrix" $\mathbf{D} = 2\sum_i \mathbf{P}^i$; if doubly filled orbitals are used, it is again defined by Eqs. (1.5), (1.6).

Roby's charge definition has the serious drawback that with its use both atoms of the molecule HCl will carry a negative resulting charge [19], which can hardly be attributed any physical meaning. This is the case because a part of the electronic charge will be assigned to both atoms due to the interatomic overlap and, therefore, one has $\|\hat{P}_{\mathrm{H}}\varphi_i\|^2 + \|\hat{P}_{\mathrm{Cl}}\varphi_i\|^2 > 1$. Mathematically this is expressed by the fact that the subspace of one atom does not represent the orthogonal complement to the subspace of another: $\hat{P}_B \neq 1 - \hat{P}_A$, or in the general, N-atomic case

$$\sum_A \hat{P}_A \neq 1 \,. \tag{2.7}$$

One would have an equality only in the absence of interatomic overlap.*

Although interatomic overlap has a large importance for our understanding of Lewis's bonding electron pairs and other chemical phenomena, for a quantitative analysis we need a tool permitting us to decompose the different quantities in a manner in which the sum of the atomic (or atomic and diatomic) components reproduces the total molecular value of the quantity considered. For that reason we have introduced [20] a formal mathematical tool called "atomic resolution of identity" which we will consider in the next chapter.

*In the absence of interatomic overlap, $S_{\rho\mu}$ and $S_{\nu\tau}$ would vanish in Eq. (2.6), if $\rho \notin A$ and $\tau \notin A$, respectively, and Eq. (2.6) would reduce to the more familiar

$$Q_A = \sum_{\mu,\rho \in A} D_{\mu\rho} S_{\rho\mu} = Tr(\mathbf{D}_{(A)}\mathbf{S}_{(A)})$$

where $\mathbf{D}_{(A)}$, $\mathbf{S}_{(A)}$ are the intraatomic blocks of the respective matrices. In that case Roby's atomic populations coincide with the Mulliken ones and their sum equals the number of electrons, $\sum_A Q_A = N$. If, in addition, the intraatomic overlap is also absent, i.e., our basis set is orthonormalized, then we have $Q_A = \sum_{\mu \in A} D_{\mu\mu}$—a direct generalization of the result we had in the Hückel case.

3

A common framework: Atomic resolution of identity

3.1 The definition

The inequality (2.7) indicates that the use of atomic projection operators is not very well suited to identify an atom in the case when interatomic overlap is present: one cannot present an arbitrary function as the sum of its projection into the different atomic subspaces. This is a peculiarity of the basis sets used in quantum chemistry—mathematicians usually use only orthonormalized basis sets. It would hardly be the best choice to follow their approach in quantum chemistry, because of the chemical significance of the interatomic overlap. Therefore in a quantum chemical framework orthonormalized basis sets usually appear only as auxiliary entities used in the derivations or numerical calculations. (It is the usual practice to use basis sets in which even the basis functions centered on the same atom are non-orthogonal, but this is only because of computational convenience and has no conceptual significance—see, however, also Section 4.3.)

For an orthonormalized basis the sum of the projection operators on the individual basis functions represents a resolution of identity, and of course that remains valid even if one at first groups the functions into subsets. That means that if an orthonormalized basis set were used, the sum of atomic projectors would be the identity operator for the given basis set. In order to facilitate our analysis, we shall introduce a *generalization* of this idea: the "atomic decomposition of identity" [20]. This permits to treat different decomposition problems (population analysis, energy components, etc.) in a common, abstract framework and the results of different Hilbert-space or 3D schemes of analysis can be obtained by a simple substitution.

We shall decompose the unit operator $\hat{I}$ into a sum of operators $\hat{\rho}_A$ which *in some sense* are assigned to the individual atoms in the molecular system:

$$\hat{I} = \sum_A \hat{\rho}_A = \sum_A \hat{\rho}_A^\dagger \tag{3.1}$$

Depending on the actual definition of the operators $\hat{\rho}_A$, different schemes of analysis arise. We do not assume, in general, the operators $\hat{\rho}_A$ to be Hermitian; the sum of all $\hat{\rho}_A^\dagger$s, is however, the same Hermitian unit operator, of course.

Applying operator $\hat{I}$ to any one-electron "ket" $|\psi\rangle$ (either MO or AO) one gets:

$$|\psi\rangle = \hat{I}|\psi\rangle = \sum_A \hat{\rho}_A|\psi\rangle \, , \tag{3.2}$$

which means that each "ket" is decomposed into a sum of components $\hat{\rho}_A|\psi\rangle$ which are assigned to the individual atoms A in the sense of the actual definition of the operators $\hat{\rho}_A$. Taking the adjoint of Eq. (3.2) we get the similar decomposition of the "bra" $\langle\psi|$:

$$\langle\psi| = \langle\psi|\hat{I} = \sum_A \langle\psi|\hat{\rho}_A^\dagger \, . \tag{3.3}$$

into a sum of the atomic components $\langle\psi|\hat{\rho}_A^\dagger$.

Analogously, a two-electron "ket" can be decomposed by introducing two identity operators $\hat{I}(1)$ and $\hat{I}(2)$—i.e., one for each electron:

$$|\Psi(1,2)\rangle = \hat{I}(1)\hat{I}(2)|\Psi(1,2)\rangle = \sum_{A,B} \hat{\rho}_A(1)\hat{\rho}_B(2)|\Psi(1,2)\rangle \, . \tag{3.4}$$

leading to a decomposition of $|\Psi(1,2)\rangle$ into a sum of monoatomic (if $A = B$) and diatomic (if $A \neq B$) components. One could, of course, take the adjoint of Eq. (3.4) to get a decomposition of the respective two-electron "bra," but we will not need it.

The two-electron "kets" we will consider represent simple products of one-electron functions $|\Psi(1,2)\rangle = |\psi(1)\psi'(2)\rangle$, where $\psi(1)$, $\psi'(2)$ may again be either AOs or MOs. Then we may write

$$|\psi(1)\psi'(2)\rangle = \sum_{A,B} |\hat{\rho}_A\psi(1)\hat{\rho}_B\psi'(2)\rangle \, . \tag{3.5}$$

If one considers the one-electron integral $\langle\varphi|\hat{l}|\psi\rangle$, where $\hat{l}$ is any linear one-electron operator and φ, ψ are arbitrary one-electron functions, then one may insert one or two identity operators decomposed according to Eq. (3.1); in other words, one may substitute either the expansion (3.2) for the "ket," or the expansion (3.2) for the "ket" *and* the expansion (3.3) for the "bra," respectively. (In the "bra" one has to write $\langle\varphi|$ instead of $\langle\psi|$, of course.) Then one gets

$$\langle\varphi|\hat{l}|\psi\rangle = \sum_A \langle\varphi|\hat{l}\hat{\rho}_A|\psi\rangle = \sum_{A,B} \langle\varphi|\hat{\rho}_A^\dagger\hat{l}\hat{\rho}_B|\psi\rangle \tag{3.6}$$

permitting to decompose the integral in question either into atomic or into mono- and diatomic components. Similarly, using (3.5) one can decompose every two-electron integral as

$$\langle\varphi(1)\varphi'(2)|\hat{g}(1,2)|\psi(1)\psi'(2)\rangle = \sum_{A,B}\langle\varphi(1)\varphi'(2)|\hat{g}(1,2)|\hat{\rho}_A\psi(1)\hat{\rho}_B\psi'(2)\rangle\,. \tag{3.7}$$

Here $\hat{g}(1,2)$ is a two-electron operator; in the most common case $\hat{g}(1,2) = 1/r_{12}$, but we shall consider other cases, too.

Of course, one could also perform the decomposition of the two-electron "bra" in (3.7) and end up with components of up to four-atom character. This would be in accord with the fact that the two-electron integrals over the basis orbitals are also up to four-center ones. However, one does not have so large *chemical* effects of three- and four-atomic character, which would correspond to such an approach.* In fact, in chemistry one has a pronounced *pairwise* interaction between the atoms, and even the aromaticity of the benzene molecule, this *par excellence* collective effect of the six-membered ring, could qualitatively be understood even by the simple Hückel theory using only two-center integrals: it is sufficient to speak about a specific interference of different mono- and diatomic effects, but there is no need to introduce multicenter energy components. This conclusion is also supported by the fact that—as will be seen in Section 9.6—the Hartree–Fock energy can be in a most natural manner presented as an *exact* sum of mono- and diatomic energy components in the framework of Bader's AIM formalism.

3.2 Different atomic operators

As discussed in Chapter 1, there are two significantly different schemes of analysis, the Hilbert-space one performed in terms of the basis functions assigned to the individual atoms and the 3D one in which the atoms are identified with (disjunct or overlapping) domains of the physical space. (Mixed treatments are also possible.) Every such scheme is uniquely determined by the definition of the respective atomic operators $\hat{\rho}_A$.

*The "natural" energy decomposition scheme of Clementi [21], based on the systematization of the one- and two-electron integrals in the different terms of the molecular energy, led to pretty large three- and four-center energy components that did not permit any reasonable chemical interpretation. This resulted in the *ab initio* energy analysis stagnating for decades.

Hilbert-space analysis:

One defines the operator $\hat{\rho}_A$ as

$$\hat{\rho}_A = \sum_{\mu \in A} |\chi_\mu\rangle\langle\widetilde{\chi}_\mu| = \sum_{\mu \in A}\sum_{\tau} |\chi_\mu\rangle(\mathbf{S}^{-1})_{\mu\tau}\langle\chi_\tau| \tag{3.8}$$

where $\widetilde{\chi}_\mu = \sum_\tau (\mathbf{S}^{-1})_{\tau\mu}\chi_\tau$ is the *biorthogonal* counterpart of orbital χ_μ. Operator $\hat{\rho}_A$ defined in this manner is nothing more than a "cut-off" operator: when applying it to a molecular orbital $|\varphi\rangle = \sum_\nu c_\nu |\chi_\nu\rangle$, one gets

$$\hat{\rho}_A|\varphi\rangle = \hat{\rho}_A \sum_\nu c_\nu|\chi_\nu\rangle = \sum_\nu c_\nu \sum_{\mu \in A} |\chi_\mu\rangle\langle\widetilde{\chi}_\mu|\chi_\nu\rangle = \sum_\nu c_\nu \sum_{\mu \in A} |\chi_\mu\rangle\delta_{\mu\nu} = \sum_{\nu \in A} c_\nu|\chi_\nu\rangle \tag{3.9}$$

i.e., the MO is cut in such a manner that only components with basis orbitals centered on atom A are conserved.

3D space analysis:

One has to decide whether or not (or to a what extent) a given point of the space belongs to a given atom. For that reason one introduces for each point of the 3D space a weight function $w_A(\vec{r})$ corresponding to the given atom A, and writes

$$\hat{\rho}_A = w_A(\vec{r}')|_{\vec{r}'=\vec{r}} \tag{3.10}$$

Here the notation $\vec{r}' = \vec{r}$ indicates that one should replace $\vec{r}'$ by $\vec{r}$ *after* the action of all the operators on the wave functions is evaluated, but before the integration over $\vec{r}$ is carried out. In this manner one can ensure that quantum mechanical operators act only on the electronic wave functions but not on the weight functions $w_A(\vec{r})$. (Similar notations are often used in the theory of density matrices.) Applying to a function $\varphi(\vec{r}) = \sum_\mu c_\nu \chi_\nu(\vec{r})$ one simply has

$$\hat{\rho}_A \varphi(\vec{r}) = w_A(\vec{r})\varphi(\vec{r}) = \sum_\nu c_\nu w_A(\vec{r})\chi_\nu(\vec{r}) \tag{3.11}$$

The weight functions assigned to different atoms should satisfy the conditions

$$w_A(\vec{r}) \geq 0 \tag{3.12}$$

and

$$\sum_A w_A(\vec{r}) = 1 \tag{3.13}$$

everywhere.

One may distinguish two types of weight functions. One pertains to Bader's "atoms in molecules" (AIM) theory [2], or to any other formalism in which the 3D space is decomposed into disjunct atomic domains Ω_A:

$$w_A(\vec{r}) = \begin{cases} 1 \text{ if } \vec{r} \in \Omega_A \,, \\ 0 \text{ otherwise} \,; \end{cases} \tag{3.14}$$

and another is the case of "fuzzy" atoms [3], i.e., division of the 3D space into atomic regions without sharp boundaries. For "fuzzy" atoms one considers weight functions providing that $w_A(\vec{r})$ is large if $\vec{r}$ is "inside" the atomic region and is small "outside" it. (Obviously the AIM condition represents a special limiting case of the "fuzzy" atoms formalism.)

Hybrid scheme:

One can consider also a scheme which combines in some sense the above two more conventional ones, and define

$$\hat{\rho}_A = \sum_{\rho} |\chi_\rho\rangle\langle\widetilde{\chi}_\rho|w_A(\vec{r}) = \sum_{\rho,\tau} |\chi_\rho\rangle S^{-1}_{\rho\tau}\langle\chi_\tau|w_A(\vec{r}) \tag{3.15}$$

In the limit of infinite basis set it approaches the 3D decomposition with the given $w_A(\vec{r})$ at the same time it permits the use of a formalism resembling the Hilbert-space one. This scheme is to some extent analogous to the projection operator formalism of Clark and Davidson [22,23].

It may be mentioned that out of the above operators $\hat{\rho}_A$ only that corresponding to the AIM scheme is a true projection operator, because in that case $w_A^2(\vec{r}) = w_A(\vec{r})$. No idempotency exists for the "fuzzy" atoms (and the hybrid scheme) as in that case $w_A^2(\vec{r}) \neq w_A(\vec{r})$; the Hilbert-space operator $\hat{\rho}_A$ (3.8) is idempotent *but is not Hermitian.* (The operator $\hat{\rho}_A^{\lambda}$ used in Section 4.3 for a Löwdin-orthogonalized basis is, however, a true projector.)

Now we are going to discuss some important formal relationships which can be obtained by using the atomic resolution of identity and the actual formulae which one obtains for their Hilbert-space realization—i.e., by substituting in them the actual form (3.8) of the operator $\hat{\rho}_A$.

3.3 Population analysis

As already noted, the distribution of the electron density is always given by an expression like Eq. (1.4) whenever a basis function constructed by using a finite basis set is used, i.e.,

$$\varrho(\vec{r}) = \sum_{\mu,\nu} D_{\mu\nu}\chi_\nu^*(\vec{r})\chi_\mu(\vec{r}) \,. \tag{3.16}$$

Integrating both sides of Eq. (3.16) over the whole space, we get the total number of electrons

$$N = \int \varrho(\vec{r})dv = \sum_{\mu,\nu} D_{\mu\nu}\int \chi_\nu^*(\vec{r})\chi_\mu(\vec{r})dv = \sum_{\mu,\nu} D_{\mu\nu}\langle\chi_\nu|\chi_\mu\rangle = \sum_{\mu,\nu} D_{\mu\nu}S_{\nu\mu} \tag{3.17}$$

An overlap integral $\langle\varphi|\psi\rangle$ is a special case of the integral $\langle\varphi|\hat{l}|\psi\rangle$, which one obtains when the operator $\hat{l}$ is a unit operator. Thus, using the results of the previous subsection, each overlap integral $S_{\mu\nu} = \langle\chi_\mu|\chi_\nu\rangle$ can be decomposed into a sum of "atomic overlap integrals"

$$S_{\mu\nu} = \langle\chi_\mu|\chi_\nu\rangle = \sum_A \langle\chi_\mu|\hat{\rho}_A\chi_\nu\rangle = \sum_A \langle\chi_\mu|\chi_\nu\rangle_A = \sum_A S^A_{\mu\nu} \,, \tag{3.18}$$

in terms of various notations. This means a decomposition of the overlap matrix **S** having the elements $S_{\mu\nu} = \langle\chi_\mu|\chi_\nu\rangle$ into a sum of partial overlap matrices* $\mathbf{S}^A$ having the elements $S^A_{\mu\nu} = \langle\chi_\mu|\chi_\nu\rangle_A = \langle\chi_\mu|\hat{\rho}_A\chi_\nu\rangle$. As is easy to see, the Hilbert-space operator $\hat{\rho}_A$ (3.8) leads to a "vertical band structure" of matrix $\mathbf{S}^A$, because $S^A_{\mu\nu} = S_{\mu\nu}$ if $\nu \in A$, and $S^A_{\mu\nu} = 0$ otherwise; in the 3D case one has simply to perform the actual integrations $S^A_{\mu\nu} = \int w_A(\vec{r})\chi_\mu^*(\vec{r})\chi_\nu(\vec{r})dv$.

Substituting the expansion (3.18) into the right-hand side of Eq. (3.17) we get the decomposition of the number of electrons into a sum of atomic "gross" populations Q_A:

$$N = \sum_A Q_A \tag{3.19}$$

where

$$Q_A = \sum_{\mu,\nu} D_{\mu\nu}\langle\chi_\nu|\hat{\rho}_A\chi_\mu\rangle = \sum_{\mu,\nu} D_{\mu\nu}\langle\chi_\nu|\chi_\mu\rangle_A = \sum_{\mu,\nu} D_{\mu\nu}S^A_{\nu\mu} \tag{3.20}$$

However, one may also apply the second form of Eq. (3.6) for the overlap integral, and then we have the expansion of the number of electrons into a double sum

$$N = \sum_{A,B} q_{AB} \tag{3.21}$$

*One has to distinguish the partial overlap matrix $\mathbf{S}^\mathbf{A}$ from the *intraatomic block* $\mathbf{S}_{(A)}$ of the original overlap matrix, which consists of those elements $S_{\mu\nu}$ for which $\mu, \nu \in A$.

where

$$q_{AB} = \sum_{\mu,\nu} D_{\mu\nu} \langle \chi_\nu | \hat{\rho}_B^\dagger \hat{\rho}_A | \chi_\mu \rangle \tag{3.22}$$

The terms q_{AA}, i.e., those terms on the right-hand side of the expansion (3.22) for which $A = B$, represent the "net" electron population of atom A, while those terms for which $A \neq B$ give the "overlap populations." Both terms "net populations" and "overlap populations" are to be understood in a generalized sense, as they are depending on the actual definition of the operators $\hat{\rho}_A$.

Between these quantities the relationship

$$Q_A = \sum_B q_{AB} \tag{3.23}$$

holds. In fact, summing both sides of (3.22) over B we have—c.f. Eq. (3.1)

$$\sum_B q_{AB} = \sum_B \sum_{\mu,\nu} D_{\mu\nu} \langle \chi_\nu | \hat{\rho}_B^\dagger \hat{\rho}_A | \chi_\mu \rangle = \sum_{\mu,\nu} D_{\mu\nu} \langle \chi_\nu | \hat{\rho}_A \chi_\mu \rangle = Q_A \tag{3.24}$$

In the case of Hilbert space analysis, one substitutes the definition (3.8) in (3.20) and (3.22), and gets the familiar expressions of Mulliken's gross, net (if $A = B$) and overlap ($A \neq B$) populations:

$$\begin{aligned} Q_A &= \sum_{\mu,\nu} D_{\mu\nu} \langle \chi_\nu | \hat{\rho}_A \chi_\mu \rangle = \sum_{\mu,\nu} D_{\mu\nu} \langle \chi_\nu | \sum_{\lambda \in A} | \chi_\lambda \rangle \langle \widetilde{\chi}_\lambda | \chi_\mu \rangle \\ &= \sum_{\mu,\nu} \sum_{\lambda \in A} D_{\mu\nu} S_{\nu\lambda} \delta_{\lambda\mu} = \sum_{\lambda \in A} \sum_{\nu} D_{\lambda\nu} S_{\nu\lambda} = \sum_{\lambda \in A} (\mathbf{DS})_{\lambda\lambda} , \end{aligned} \tag{3.25}$$

and

$$\begin{aligned} q_{AB} &= \sum_{\mu,\nu} D_{\mu\nu} \langle \chi_\nu | \hat{\rho}_B^\dagger \hat{\rho}_A | \chi_\mu \rangle = \sum_{\mu,\nu} D_{\mu\nu} \langle \chi_\nu | \sum_{\tau \in B} | \widetilde{\chi}_\tau \rangle \langle \chi_\tau | \sum_{\lambda \in A} | \chi_\lambda \rangle \langle \widetilde{\chi}_\lambda | \chi_\mu \rangle \\ &= \sum_{\mu,\nu} \sum_{\lambda \in A} \sum_{\tau \in B} D_{\mu\nu} \delta_{\nu\tau} S_{\tau\lambda} \delta_{\lambda\mu} = \sum_{\lambda \in A} \sum_{\tau \in B} D_{\lambda\tau} S_{\tau\lambda} . \end{aligned} \tag{3.26}$$

The numerical results given by the Mulliken population analysis often exhibit a strong basis dependence. Nonetheless, the above results indicate that the Mulliken population analysis has a privileged conceptual importance in the framework of the Hilbert-space analysis, because it is that one which is consistent with the internal mathematical structure of the finite basis LCAO theory, i.e., with the use of atom-centered basis orbitals (c.f. also [5,24,26,27]).

3.4 Decomposition of electron density and of exchange density

Atomic electron densities

It may be of interest to present not only the overall number of electrons as a sum of atomic (or atomic and overlap) populations, but also to decompose the electron density function $\rho(\vec{r})$ itself into a sum of terms assigned to the individual atoms or pairs of atoms, and get atomic (or net atomic and overlap) densities. If one uses the 3D analysis, then this can easily be accomplished by multiplying the electron density function with one or two "atomic resolutions of identity" written down in terms of operators (3.10). In the Hilbert space formalism, simple inspection of the electron density written in the form of Eq. (3.16) suggests that the adequate partitioning would be (e.g., [28]) the decomposition

$$\varrho(\vec{r}) = \sum_A \varrho_A(\vec{r}) \,, \tag{3.27}$$

with the "atomic densities"

$$\varrho_A(\vec{r}) = \sum_{\mu\in A}\sum_{\nu} D_{\mu\nu}\chi_\nu^*(\vec{r})\chi_\mu(\vec{r}) \,. \tag{3.28}$$

which integrate to the respective Mulliken's gross atomic population Q_A. One can analogously define the net and overlap densities as

$$\varrho_{AA}(\vec{r}) = \sum_{\mu,\nu\in A} D_{\mu\nu}\chi_\nu^*(\vec{r})\chi_\mu(\vec{r}) \,, \qquad \varrho_{AB}(\vec{r}) = \sum_{\mu\in A}\sum_{\nu\in B} D_{\mu\nu}\chi_\nu^*(\vec{r})\chi_\mu(\vec{r}) \,. \tag{3.29}$$

again integrating to the net atomic populations q_{AA} and overlap populations q_{AB}, respectively, and satisfying the requirement $\varrho_A(\vec{r}) = \sum_B \varrho_{AB}(\vec{r})$.

In the Hilbert space formalism one cannot apply the "atomic resolution of identity" *directly* to the electron density, because the atomic operators (3.8) are in terms of Dirac's "bras" and "kets" while the function $\rho(\vec{r})$ is defined in the 3D physical space. It may be, however, worthwhile to demonstrate that the decompositions (3.27)–(3.29) can also be derived by using the atomic operators $\hat{\rho}_A$.

The electron density in the point $\vec{r}'$ can be written as the expectation value of the operator sum $\sum_i \delta(\vec{r}_i - \vec{r}')$, where $\delta(\vec{r}_i - \vec{r}')$ is the Dirac delta function*

$$\varrho(\vec{r}') = \frac{\langle\Psi|\sum_i \delta(\vec{r}_i - \vec{r}')|\Psi\rangle}{\langle\Psi|\Psi\rangle} \tag{3.30}$$

Now, this can be treated as the expectation value of any symmetric sum of one-electron operators [5], and will lead to some sum of one-electron integrals of the form $\langle\varphi(\vec{r}_i)|\delta(\vec{r}_i - \vec{r}')|\psi(\vec{r}_i)\rangle$, with one-electron functions φ, ψ and coefficients depending on the actual form of the wave function. For instance, if Ψ is a single determinant wave function built up of the orthonormalized spin-orbitals $\psi_i = \psi_i(\vec{r}, s) = \varphi_i^\sigma(\vec{r})\sigma(s)$, where $\varphi_i^\sigma(\vec{r})$ and $\sigma(s)$ denote the spatial and spin parts of ψ_i, respectively ($\sigma = \alpha$ or β), then

$$\begin{aligned}\varrho(\vec{r}') &= \sum_i \langle\psi_i(\vec{r}, s)|\delta(\vec{r} - \vec{r}')|\psi_i(\vec{r}, s)\rangle \\ &= \sum_{i=1}^{N_\alpha} \langle\varphi_i^\alpha|\delta(\vec{r} - \vec{r}')|\varphi_i^\alpha\rangle + \sum_{i=1}^{N_\beta} \langle\varphi_i^\beta|\delta(\vec{r} - \vec{r}')|\varphi_i^\beta\rangle\end{aligned} \tag{3.31}$$

where N_α and N_β are the numbers of occupied orbitals (and electrons) of the respective spin and the integrals are over the unprimed variables.

Now, one can apply the operator $\hat{\rho}_A$ as defined in (3.8) to the "kets" on the right-hand side. Inserting an "atomic resolution of identity" to the "kets" and substituting the LCAO expansions of the orbitals

$$|\varphi_i^\sigma\rangle = \sum_\mu c_\mu^{\sigma i}|\chi_\mu\rangle \ , \qquad (\sigma = \alpha \text{ or } \beta) \tag{3.32}$$

(and similarly for "bras"), a simple derivation quite analogous to that given above gives the decomposition of $\varrho(\vec{r})$ into a sum of "atomic densities," each integrating to the respective Mulliken's gross atomic population:

$$\begin{aligned}\varrho(\vec{r}') &= \sum_A \sum_{i=1}^{N_\alpha} \sum_{\mu,\nu} c_\mu^{\alpha i} c_\nu^{\alpha i*} \langle\chi_\nu|\delta(\vec{r} - \vec{r}')\hat{\rho}_A|\chi_\mu\rangle \\ &+ \sum_A \sum_{i=1}^{N_\beta} \sum_{\mu,\nu} c_\mu^{\beta i} c_\nu^{\beta i*} \langle\chi_\nu|\delta(\vec{r} - \vec{r}')\hat{\rho}_A|\chi_\mu\rangle \\ &= \sum_A \sum_{\mu\in A} \sum_\nu \left[P_{\mu\nu}^\alpha \langle\chi_\nu|\delta(\vec{r} - \vec{r}')|\chi_\mu\rangle + P_{\mu\nu}^\beta \langle\chi_\nu|\delta(\vec{r} - \vec{r}')|\chi_\mu\rangle\right]\end{aligned} \tag{3.33}$$

*One usually defines the electron density on the basis that the probability of finding the i-th electron in the vicinity of point $\vec{r}$ can be obtained by integrating the square of the absolute value of the wave function over the coordinates of all electrons except the i-th, and then one has to sum over all the electrons. Eq. (3.30) is simply the mathematical formulation of this procedure.

$$= \sum_{A} \sum_{\mu \in A} \sum_{\nu} D_{\mu\nu} \chi_\nu^*(\vec{r}\,') \chi_\mu(\vec{r}\,') = \sum_{A} \varrho_A(\vec{r}\,'),$$

because $\hat{\rho}_A|\chi_\mu\rangle$ survives only if $\mu \in A$ and $\langle\chi_\nu|\delta(\vec{r}-\vec{r}\,')|\chi_\mu\rangle = \chi_\nu^*(\vec{r}\,')\chi_\mu(\vec{r}\,')$ according to the properties of the delta function.

In these formulae we have introduced the "density matrices"

$$\mathbf{P}^\alpha = \sum_{i=1}^{N_\alpha} \mathbf{c}^{\alpha i}\mathbf{c}^{\alpha i\dagger}, \qquad \mathbf{P}^\beta = \sum_{i=1}^{N_\beta} \mathbf{c}^{\beta i}\mathbf{c}^{\beta i\dagger}, \tag{3.34}$$

having the elements

$$P_{\mu\nu}^\alpha = \sum_{i=1}^{N_\alpha} c_\mu^{\alpha i} c_\nu^{\alpha i *}, \qquad P_{\mu\nu}^\beta = \sum_{i=1}^{N_\beta} c_\mu^{\beta i} c_\nu^{\beta i *}, \tag{3.35}$$

respectively, and the "spinless density matrix"

$$\mathbf{D} = \mathbf{P}^\alpha + \mathbf{P}^\beta. \tag{3.36}$$

These notations will be used throughout the book.

Similarly, applying a resolution of identity to the "bras", too, we get the net atomic densities and overlap densities, again integrating to the respective net and overlap populations. It is important to stress that for a pair of orthogonal orbitals their contribution to the respective atomic or interatomic population vanishes; this, however, does not hold for their overlap density: simply it has both positive and negative domains which cancel at the final integration.*

In the papers [29,30] we have considered an alternative definition (or approximation) of the atomic densities $\varrho_A(\vec{r})$, which can be obtained from Eq. (3.28) by replacing each function $\chi_\nu^*(\vec{r})$ not centered on the given atom A by its projection on the atomic subspace. The functions obtained in that manner also integrate to the respective Mulliken population of the atoms, but the sum of these "projected" atomic densities does not strictly equal the overall electron density $\varrho(\vec{r})$.

Exchange density

Similarly to the electron density $\varrho(\vec{r})$ giving the probability of finding an electron in the vicinity of the point $\vec{r}$, one may also define the *pair density* $\varrho_2(\vec{r},\vec{r}\,')$ giving the probability of finding one electron in the vicinity of the

*In such a case one cannot use the "normalized overlap density" (page 134 of [31]), as it would mean division by zero.

point $\vec{r}$ and another in the vicinity of the point $\vec{r}'$. It is easy to see (c.f. [5]) that it is given by the expectation value of the operator

$$\sum_{\substack{i,j \\ (i\neq j)}} \delta(\vec{r}_i-\vec{r})\delta(\vec{r}_j-\vec{r}') = \sum_{i<j}\left[\delta(\vec{r}_i-\vec{r})\delta(\vec{r}_j-\vec{r}')+\delta(\vec{r}_j-\vec{r})\delta(\vec{r}_i-\vec{r}')\right] . \tag{3.37}$$

In the single determinant case this expectation value can be computed by using the standard formulae for the symmetric sums of two-electron operators (Slater rules), giving

$$\begin{aligned}\varrho_2(\vec{r},\vec{r}') = {} & \frac{1}{2}\sum_{i,j=1}^{N_\alpha}\left[\langle\varphi_i^\alpha(\vec{r}_1)\varphi_j^\alpha(\vec{r}_2)|\hat{g}(1,2)|\varphi_i^\alpha(\vec{r}_1)\varphi_j^\alpha(\vec{r}_2)\rangle\right. \\ & \left. - \langle\varphi_i^\alpha(\vec{r}_1)\varphi_j^\alpha(\vec{r}_2)|\hat{g}(1,2)|\varphi_j^\alpha(\vec{r}_1)\varphi_i^\alpha(\vec{r}_2)\rangle\right] \\ & + \frac{1}{2}\sum_{i,j=1}^{N_\beta}\left[\langle\varphi_i^\beta(\vec{r}_1)\varphi_j^\beta(\vec{r}_2)|\hat{g}(1,2)|\varphi_i^\beta(\vec{r}_1)\varphi_j^\beta(\vec{r}_2)\rangle\right. \\ & \left. - \langle\varphi_i^\beta(\vec{r}_1)\varphi_j^\beta(\vec{r}_2)|\hat{g}(1,2)|\varphi_j^\beta(\vec{r}_1)\varphi_i^\beta(\vec{r}_2)\rangle\right] \\ & + \sum_{i=1}^{N_\alpha}\sum_{j=1}^{N_\beta}\langle\varphi_i^\alpha(\vec{r}_1)\varphi_j^\beta(\vec{r}_2)|\hat{g}(1,2)|\varphi_i^\alpha(\vec{r}_1)\varphi_j^\beta(\vec{r}_2)\rangle ,\end{aligned} \tag{3.38}$$

where

$$\hat{g}(1,2) = \delta(\vec{r}_1-\vec{r})\delta(\vec{r}_2-\vec{r}')+\delta(\vec{r}_2-\vec{r})\delta(\vec{r}_1-\vec{r}') . \tag{3.39}$$

Substituting (3.39) into (3.38), performing the integrations by taking into account the properties of the delta-function and collecting the terms yielding the electron density

$$\varrho(\vec{r}) = \sum_{i=1}^{N_\alpha}|\varphi_i^\alpha(\vec{r})|^2+\sum_{i=1}^{N_\beta}|\varphi_i^\beta(\vec{r})|^2 , \tag{3.40}$$

c.f. Eq. (1.3), we get after some manipulations (interchange of summation indices, etc.) the expression of the pair density

$$\varrho_2(\vec{r},\vec{r}') = \varrho(\vec{r})\varrho(\vec{r}')-\varrho_2^x(\vec{r},\vec{r}') , \tag{3.41}$$

where the *exchange density* $\varrho_2^x(\vec{r},\vec{r}')$ is defined as

$$\begin{aligned}\varrho_2^x(\vec{r},\vec{r}') = {} & \sum_{i,j=1}^{N_\alpha}\varphi_i^{\alpha*}(\vec{r})\varphi_i^\alpha(\vec{r}')\varphi_j^{\alpha*}(\vec{r}')\varphi_j^\alpha(\vec{r}) \\ & + \sum_{i,j=1}^{N_\beta}\varphi_i^{\beta*}(\vec{r})\varphi_i^\beta(\vec{r}')\varphi_j^{\beta*}(\vec{r}')\varphi_j^\beta(\vec{r}) .\end{aligned} \tag{3.42}$$

Substituting the LCAO expansions (3.32) of the orbitals and using the definitions (3.35), this can be rewritten also as

$$\varrho_2^x(\vec{r},\vec{r}\,') = \sum_{\mu,\nu,\rho,\tau} \left(P^{\alpha}_{\nu\mu} P^{\alpha}_{\tau\rho} + P^{\beta}_{\nu\mu} P^{\beta}_{\tau\rho} \right) \chi^*_{\mu}(\vec{r}) \chi_{\nu}(\vec{r}\,') \chi^*_{\rho}(\vec{r}\,') \chi_{\tau}(\vec{r}) \tag{3.43}$$

It is easy to see by integrating Eq. (3.42) that the exchange density integrates to the number of electrons (the molecular orbitals are orthonormalized):

$$\iint \varrho_2^x(\vec{r},\vec{r}\,') dv dv' = N_\alpha + N_\beta = N \,. \tag{3.44}$$

The same result can be obtained by using Eq. (3.43), too: integrating both sides we get

$$\iint \varrho_2^x(\vec{r},\vec{r}\,') dv dv' = \sum_{\nu,\tau} \left[(\mathbf{P}^{\alpha}\mathbf{S})_{\nu\tau} (\mathbf{P}^{\alpha}\mathbf{S})_{\tau\nu} + (\mathbf{P}^{\beta}\mathbf{S})_{\nu\tau} (\mathbf{P}^{\beta}\mathbf{S})_{\tau\nu} \right] \tag{3.45}$$

Utilizing the idempotency of matrices $\mathbf{P}^{\sigma}\mathbf{S}$ and the known property $\mathrm{Tr}(\mathbf{P}^{\sigma}\mathbf{S}) = N_\sigma$ we arrive at (3.44).

In the framework of a 3D analysis, it does not represent a problem to decompose the expressions (3.42), (3.43) of the exchange density in the physical space, or its integral (3.44), into one- and diatomic components. One has simply to insert two atomic decompositions of identity (3.1)—one for $\vec{r}$, another for $\vec{r}\,'$—in terms of atomic operators (3.10) utilizing the 3D weight functions. In the case of a Hilbert-space analysis, we have to proceed similarly to the approach used above for obtaining atomic densities.

Inspection of the expressions (3.38) and (3.42) indicates that the exchange density is given by the sum of integrals entering with sign minus in the right-hand side of (3.38):

$$\begin{aligned} \varrho_2^x(\vec{r},\vec{r}\,') &= \frac{1}{2} \sum_{i,j=1}^{N_\alpha} \langle \varphi_i^{\alpha}(\vec{r}_1) \varphi_j^{\alpha}(\vec{r}_2) | \hat{g}(1,2) | \varphi_j^{\alpha}(\vec{r}_1) \varphi_i^{\alpha}(\vec{r}_2) \rangle \\ &+ \frac{1}{2} \sum_{i,j=1}^{N_\beta} \langle \varphi_i^{\beta}(\vec{r}_1) \varphi_j^{\beta}(\vec{r}_2) | \hat{g}(1,2) | \varphi_j^{\beta}(\vec{r}_1) \varphi_i^{\beta}(\vec{r}_2) \rangle \,. \end{aligned} \tag{3.46}$$

We can apply the decomposition (3.7) to every integral and get after substituting the LCAO expansions (3.32) of the orbitals and utilizing the properties of the delta-function to evaluate the integrals with the operators $\hat{g}(1,2)$ defined in (3.39) the expression

$$\varrho_2^x(\vec{r},\vec{r}\,') = \sum_{A,B} \sum_{\tau,\rho} \sum_{\mu \in A} \sum_{\nu \in B} \left(P^{\alpha}_{\mu\tau} P^{\alpha}_{\nu\rho} + P^{\beta}_{\mu\tau} P^{\beta}_{\nu\rho} \right) \chi^*_{\tau}(\vec{r}) \chi_{\mu}(\vec{r}\,') \chi^*_{\rho}(\vec{r}\,') \chi_{\nu}(\vec{r}) \,, \tag{3.47}$$

where the definitions (3.35) of the P-matrix elements have also been taken into account. (Some summation indices had again to be interchanged.) Comparison with the expression (3.43) indicates that we assign the different terms to the individual atoms or pairs of atoms according to the basis functions originating from the "kets" of the integrals in (3.46)—as it should be. That means that we assign to each pair of atoms A and B the well-defined component of the exchange density

$$\sum_{\mu\in A}\sum_{\nu\in B}\sum_{\tau,\rho}\left(P^{\alpha}_{\mu\tau}P^{\alpha}_{\nu\rho}+P^{\beta}_{\mu\tau}P^{\beta}_{\nu\rho}\right)\chi^{*}_{\tau}(\vec{r})\chi_{\mu}(\vec{r}')\chi^{*}_{\rho}(\vec{r}')\chi_{\nu}(\vec{r})+(A\leftrightarrow B)\,, \tag{3.48}$$

where $(A \leftrightarrow B)$ denotes the terms with A and B interchanged. This is the exchange density analogue of the atomic density expression (3.28).

Integrating both sides of Eq. (3.47) we get the decomposition of the integral (3.45) into one- and two-center components

$$\iint \varrho^{x}_{2}(\vec{r},\vec{r}')dvdv' = \sum_{A,B}\sum_{\mu\in A}\sum_{\nu\in B}\left[(\mathbf{P}^{\alpha}\mathbf{S})_{\mu\nu}(\mathbf{P}^{\alpha}\mathbf{S})_{\nu\mu}+(\mathbf{P}^{\beta}\mathbf{S})_{\mu\nu}(\mathbf{P}^{\beta}\mathbf{S})_{\nu\mu}\right]\,, \tag{3.49}$$

where the two-center components are in a direct relationship with the *bond order indices* to be discussed in Section 6.3.

4

Analysis of the first-order density in Hilbert space

4.1 The LCAO representation of the first-order density matrix

The electron density $\varrho(\vec{r})$ can be considered the "diagonal element" of a more complex entity, the "spinless first-order density matrix" $\varrho_1(\vec{r};\vec{r}')$.* The name "density matrix" is somewhat conventional, as it has continuous "rows" and "columns" $\vec{r}$ and $\vec{r}'$, respectively. The "diagonal element" means that we substitute $\vec{r}' = \vec{r}$ into $\varrho_1(\vec{r};\vec{r}')$:

$$\varrho(\vec{r}) = \varrho_1(\vec{r};\vec{r}) \tag{4.1}$$

If in terms of a (finite or infinite) basis of one-electron functions $\{\chi_\mu\}$ the electron density has an expansion (3.16), then the spinless first-order density matrix can be expressed with the aid of the same matrix elements $D_{\mu\nu}$ as

$$\varrho_1(\vec{r};\vec{r}') = \sum_{\mu,\nu} D_{\mu\nu}\chi_\nu^*(\vec{r}')\chi_\mu(\vec{r}) \,. \tag{4.2}$$

where $\mathbf{D}$ is a matrix in the usual sense, i.e., its elements $D_{\mu\nu}$ depend on two discrete variables (the row and column indices). The equality (4.1) connects this expression with the expansion (3.16) of the electron density.

The spinless first-order density matrix is a sum of the first-order (spatial) density matrices corresponding to spins α and β. Using Eq. (3.36), we have

$$\varrho_1(\vec{r};\vec{r}') = \sum_{\mu,\nu}\left(P^\alpha_{\mu\nu} + P^\beta_{\mu\nu}\right)\chi_\nu^*(\vec{r}')\chi_\mu(\vec{r}) \,. \tag{4.3}$$

The importance of the spinless first-order density matrix $\varrho_1(\vec{r};\vec{r}')$ lies in the fact that the expectation value $\langle\hat{L}\rangle$ of any symmetric sum $\hat{L} = \sum_i \hat{l}(i)$ of spinless one-electron operators equals [31]

$$\langle\hat{L}\rangle = \int \hat{l}\,\varrho_1(\vec{r};\vec{r}')\Big|_{\vec{r}'=\vec{r}} dv \,, \tag{4.4}$$

*Part of this section is reprinted from my paper [27] with permission from Elsevier.

where the notation $\vec{r}' = \vec{r}$ indicates that one should replace $\vec{r}'$ by $\vec{r}$ *after* the action of operator $\hat{l}$ is evaluated, but before the integration over $\vec{r}$ is carried out. This means that $\hat{l}$ does not act on the quantities depending on the primed variable $\vec{r}'$. This is important, e.g., if $\hat{l}$ contains differentiation, as does the kinetic energy operator.†

If one also considers explicitly the spin variables, then one obtains the full (spin-dependent) first-order density matrix $\varrho_1(\vec{r},s;\vec{r}',s')$ or $\varrho_1(\mathbf{x};\mathbf{x}')$, where $\mathbf{x}$ stands for the collection of the spatial and spin coordinates $\vec{r}$ and s. It is defined as [31]

$$\varrho_1(\mathbf{x}_1;\mathbf{x}_1') = N\int\ldots\int \Psi^*(\mathbf{x}_1',\mathbf{x}_2,\ldots,\mathbf{x}_N)\Psi(\mathbf{x}_1,\mathbf{x}_2,\ldots,\mathbf{x}_N)d\mathbf{x}_2 d\mathbf{x}_3\ldots d\mathbf{x}_N\,, \tag{4.5}$$

where the subscripts "1" of the independent variables indicate that they originate from the coordinates of the first electron in the arguments of the wave function Ψ, according to which *no integration* is performed. (Integration over $\mathbf{x}_i$ includes summation over the spin s_i, too.) By introducing the notation $\chi_\mu^\sigma(\vec{r},s) = \chi_\mu(\vec{r})\sigma(s)$ for the *spin-orbitals* ($\sigma = \alpha$ or β), one gets an expansion of $\varrho_1(\mathbf{x}_1;\mathbf{x}_1')$ in terms of their products, representing a straightforward generalization of (4.3):

$$\varrho_1(\mathbf{x};\mathbf{x}') = \sum_{\mu,\nu,\sigma,\sigma'} P_{\mu\nu}^{\sigma\sigma'}\chi_\nu^{\sigma'*}(\vec{r}',s')\chi_\mu^\sigma(\vec{r},s)\,. \tag{4.6}$$

For wave functions representing eigenfunctions of the operator $\hat{S}_z$ (the number of α and β spins does not vary in different terms of the wave function), matrix $\mathbf{P}$ in Eq. (4.6) is a direct sum of the "diagonal" blocks $\mathbf{P}^{\alpha\alpha} = \mathbf{P}^\alpha$ and $\mathbf{P}^{\beta\beta} = \mathbf{P}^\beta$, while the "off-diagonal" blocks $\mathbf{P}^{\alpha\beta}$ and $\mathbf{P}^{\beta\alpha}$ vanish [31]

$$\mathbf{P} = \begin{pmatrix} \mathbf{P}^\alpha & \mathbf{0} \\ \mathbf{0} & \mathbf{P}^\beta \end{pmatrix}. \tag{4.7}$$

Thus one gets the expression

$$\varrho_1(\mathbf{x};\mathbf{x}') = \sum_{\mu,\nu}\left[P_{\mu\nu}^\alpha\chi_\nu^*(\vec{r}')\chi_\mu(\vec{r})\alpha^*(s')\alpha(s) + P_{\mu\nu}^\beta\chi_\nu^*(\vec{r}')\chi_\mu(\vec{r})\beta^*(s')\beta(s)\right], \tag{4.8}$$

†Eq. (4.4) represents a generalization of the standard expression $\langle\hat{l}\rangle = \int\psi^*(\vec{r})\hat{l}\psi(\vec{r})dv$ of the expectation value for a one-electron system with a normalized wave function ψ. In this case one has to put $\varrho_1(\vec{r};\vec{r}') = \psi^*(\vec{r}')\psi(\vec{r})$ as a special case of the expansion (4.2) and recovers $\langle\hat{l}\rangle$ by using formula (4.4). (This also explains why the complex conjugation in the expansion (4.2) should be at the functions depending on the primed variable $\vec{r}'$.)

which reduces to (4.3) after performing a summation over the spin variables.

Alternatively, by introducing a combined set of spin-orbitals $\{\chi_\mu(\mathbf{x})\}$ as the union of the sets $\{\chi_\mu^\alpha(\vec{r},s)\}$ and $\{\chi_\mu^\beta(\vec{r},s)\}$, one can simply write

$$\varrho_1(\mathbf{x};\mathbf{x}') = \sum_{\mu,\nu} P_{\mu\nu}\chi_\nu^*(\mathbf{x}')\chi_\mu(\mathbf{x})\,. \tag{4.9}$$

Of course, the summation indices μ, ν now run over both sets $\{\chi_\mu^\alpha\}$ and $\{\chi_\mu^\beta\}$, but—in accord with (4.7)—the matrix elements $P_{\mu\nu}$ are non-zero only for spin-orbitals of the same spin.

The first-order density matrix $\varrho_1(\mathbf{x};\mathbf{x}')$ is the kernel of the integral-operator $\hat{\varrho}$ called "density operator"; it acts on the arbitrary spin-orbital $\psi(\mathbf{x})$ as

$$\hat{\varrho}\psi(\mathbf{x}) = \int \varrho_1(\mathbf{x};\mathbf{x}')\psi(\mathbf{x}')d\mathbf{x}' = \sum_{\mu,\nu} P_{\mu\nu}\chi_\mu(\mathbf{x})\int \chi_\nu^*(\mathbf{x}')\psi(\mathbf{x}')d\mathbf{x}'\,. \tag{4.10}$$

In the case of a single determinant wave function, operator $\hat{\varrho}$ performs the *projection* of an arbitrary spin-orbital on the subspace of occupied spin-orbitals.*

There is a significant difference between the mathematical properties of the LCAO representation of the first-order density matrix, depending on whether or not the basis used is orthonormalized. Thus, in the case when the basis $\{\chi_\mu\}$ is *orthonormalized,* the matrix **P** has a threefold significance [27].

(i) Its elements represent the coefficients in the LCAO-type expansion (4.9) of the first-order density matrix $\varrho_1(\mathbf{x};\mathbf{x}')$.

(ii) The matrix **P** gives the LCAO representation of the density operator $\hat{\varrho}$ in the given basis. This means that if the LCAO coefficients of the arbitrary spin-orbital $\psi = \psi(\mathbf{x})$ form a column vector **q**, then the function resulting from the operation $\hat{\varrho}\psi$ has a vector of LCAO expansion coefficients equal to **Pq**. In fact, if $\psi(\mathbf{x}) = \sum_\rho q_\rho\chi_\rho(\mathbf{x})$, then

$$\begin{aligned}\hat{\varrho}\psi(\mathbf{x}) &= \int \varrho_1(\mathbf{x};\mathbf{x}')\psi(\mathbf{x}')d\mathbf{x}' = \sum_{\mu,\nu} P_{\mu\nu}\chi_\mu(\mathbf{x})\int \chi_\nu^*(\mathbf{x}')\sum_\rho q_\rho\chi_\rho(\mathbf{x}')d\mathbf{x}' \\ &= \sum_{\mu,\nu,\rho} P_{\mu\nu}\delta_{\nu\rho}q_\rho\chi_\mu(\mathbf{x}) = \sum_\mu(\mathbf{Pq})_\mu\chi_\mu(\mathbf{x})\,.\end{aligned} \tag{4.11}$$

*We note in this context that—according to the expansion at the right-hand side of (4.10)—operator $\hat{\varrho}$ can be written in terms of the "bra" and "ket" notations as

$$\hat{\varrho} = \sum_{\mu,\nu} |\chi_\mu\rangle P_{\mu\nu}\langle\chi_\nu|\,.$$

(iii) The elements of the matrix **P** give the integrals of the density operator $\hat{\varrho}$ over the basis spin-orbitals χ:

$$\langle \chi_\lambda | \hat{\varrho} | \chi_\tau \rangle = P_{\lambda\tau} \,. \tag{4.12}$$

In fact, we have

$$\begin{aligned}\langle \chi_\lambda | \hat{\varrho} | \chi_\tau \rangle &= \int \chi_\lambda^*(\mathbf{x}) \hat{\varrho} \chi_\tau(\mathbf{x}) d\mathbf{x} = \int \chi_\lambda^*(\mathbf{x}) \int \varrho_1(\mathbf{x};\mathbf{x}') \chi_\tau(\mathbf{x}') d\mathbf{x}' \\ &= \sum_{\mu,\nu} P_{\mu\nu} \iint \chi_\lambda^*(\mathbf{x}) \chi_\mu(\mathbf{x}) \chi_\nu^*(\mathbf{x}') \chi_\tau(\mathbf{x}') d\mathbf{x} d\mathbf{x}' = \sum_{\mu,\nu} P_{\mu\nu} \delta_{\lambda\mu} \delta_{\nu\tau} = P_{\lambda\tau} \,.\end{aligned} \tag{4.13}$$

Now, if the basis set is not orthogonal—which is practically the case for all actual calculations—and the overlap integrals form the matrix **S** with the elements $S_{\mu\nu} \neq \delta_{\mu\nu}$, then instead of the single matrix **P**, there appear three different matrices in the above three cases [27]:

(i) the LCAO expansion still has the general form of Eq. (4.9) and contains the elements of the matrix **P**;

(ii) the matrix describing the action of $\hat{\varrho}$ becomes **PS** because, if the basis is not orthogonal, one has to write $S_{\nu\rho}$ instead of $\delta_{\nu\rho}$ in Equation (4.11) above;

(iii) the integrals of $\hat{\varrho}$ over the basis orbitals are given as

$$\langle \chi_\lambda | \hat{\varrho} | \chi_\tau \rangle = (\mathbf{SPS})_{\lambda\tau} \,, \tag{4.14}$$

i.e., by the elements of the matrix **SPS**, because in Eq. (4.13) one has to write $S_{\lambda\mu}$ and $S_{\nu\tau}$ instead of $\delta_{\lambda\mu}$ and $\delta_{\nu\tau}$, respectively.

The relationship between the matrices appearing in cases (ii) and (iii) is a characteristic example for the behavior of the linear operators with respect to non-orthogonal basis sets [26]. In fact, let us consider an arbitrary linear operator $\hat{L}$, and form the matrix **L**, the elements of which are the integrals of $\hat{L}$ over the basis orbitals: $L_{\mu\nu} = \langle \chi_\mu | \hat{L} | \chi_\nu \rangle$. If $\hat{L}$ is a Hermitian operator, the matrix **L** is also Hermitian, irrespective of whether the basis orbitals are orthonormalized or not. However, only for an orthonormalized basis set is the matrix **L** the proper matrix representation of the linear transformation induced by the operator $\hat{L}$. To see this, let us consider the effect of $\hat{L}$ on an arbitrary orbital ψ with the LCAO coefficients forming the column vector **q**:

$$\hat{L}\psi = \hat{L} \sum_\mu q_\mu \chi_\mu = \sum_\mu q_\mu \hat{L} \chi_\mu = \sum_\mu q_\mu \sum_\nu T_{\nu\mu} \chi_\nu = \sum_\nu \Big(\sum_\mu T_{\nu\mu} q_\mu \Big) \chi_\nu \,. \tag{4.15}$$

Thus we have to find the matrix **T**, the elements of which give the LCAO expansion of $\hat{L}\chi_\mu$:

$$\hat{L}\chi_\mu = \sum_\nu T_{\nu\mu} \chi_\nu \,. \tag{4.16}$$

By multiplying (4.16) by χ_ρ^* and integrating we have

$$L_{\rho\mu} = \sum_\nu T_{\nu\mu} S_{\rho\nu} = (\mathbf{ST})_{\rho\mu} , \tag{4.17}$$

which means that

$$\mathbf{T} = \mathbf{S}^{-1}\mathbf{L} . \tag{4.18}$$

This result indicates that for a non-orthogonal basis set the matrix $\mathbf{T}$ describing the linear transformation induced by the operator $\hat{L}$ can be obtained from the matrix $\mathbf{L}$ of the integrals of this operator over the non-orthogonal basis orbitals by multiplying from the left with the inverse overlap matrix $\mathbf{S}^{-1}$. In the present case $\hat{L} = \hat{\varrho}$, its integrals form matrix $\mathbf{L} = \mathbf{SPS}$, as discussed in case (iii) above. Then the transformation matrix $\mathbf{T}$ describing the effect of $\hat{\varrho}$ in the non-orthogonal LCAO framework can be obtained by multiplying with $\mathbf{S}^{-1}$ from the left, yielding matrix $\mathbf{T} = \mathbf{PS}$, in full accord with that described under point (ii) above.

A characteristic feature of non-orthogonal basis sets is that the matrix $\mathbf{T}$ describing a linear transformation induced by a Hermitian operator $\hat{L}$ is generally not Hermitian, as can be seen from Eq. (4.18): both $\mathbf{S}^{-1}$ and $\mathbf{L}$ are Hermitian but their product is usually not. In other words, Hermitian operators are represented by non-Hermitian matrices in non-orthogonal basis sets. However, for these non-Hermitian matrices the eigenvalue equations of type $\mathbf{Tq}_i = \varepsilon_i \mathbf{q}_i$ have all the properties which are characteristic of Hermitian matrices in orthonormalized basis sets: the eigenvalues are real and the eigenvectors form an orthogonal set spanning the whole linear space defined by the original basis orbitals. This can be proved either by turning to an auxiliary orthonormalized basis or simply by observing that multiplying with $\mathbf{S}$ from left, the eigenvalue equation $\mathbf{Tq}_i = \varepsilon_i \mathbf{q}_i$ can be transformed to the generalized eigenvalue problem $\mathbf{Lq}_i = \varepsilon_i \mathbf{Sq}_i$ with Hermitian matrices $\mathbf{L}$ and $\mathbf{S}$. (We have $\mathbf{L} = \mathbf{ST}$ according to Eq. (4.18), and $\mathbf{L}$ is Hermitian by its definition above.)

These results can be considered as a special case (or inversion) of a general theorem, according to which if a non-symmetric matrix $\mathbf{X}$ has a whole set of eigenvectors and all its eigenvalues are real, then its eigenvalue equation is equivalent to a generalized eigenvalue equation of a Hermitian matrix with a properly selected overlap (metric) matrix $\mathbf{S}$. In fact, let us have

$$\mathbf{XC} = \mathbf{C\Lambda} ; \quad \mathbf{\Lambda} = diag\{\lambda_1, \lambda_2, \ldots, \lambda_N\} ; \quad \lambda_i^* = \lambda_i, \ \forall i . \tag{4.19}$$

Then choosing $\mathbf{S} = (\mathbf{C}^{-1})^{\dagger}\mathbf{C}^{-1}$, which is Hermitian by its definition and can easily be seen to be positive definite,* and defining $\mathbf{H} = \mathbf{SX}$, we get

$$\mathbf{HC} = \mathbf{SC\Lambda} . \tag{4.20}$$

Furthermore, as $\mathbf{SC} = (\mathbf{C}^{-1})^{\dagger}$, we can write $\mathbf{H} = (\mathbf{C}^{-1})^{\dagger}\mathbf{\Lambda}\mathbf{C}^{-1}$, which shows that $\mathbf{H}$ is indeed Hermitian because all the eigenvalues λ_i are real, and thus $\mathbf{\Lambda}^{\dagger} = \mathbf{\Lambda}$.

Now, although this might be considered a question of nomenclature only, we think [27] that of the above three matrices **P**, **PS**, and **SPS**, the non-Hermitian matrix **PS** discussed in case (ii) is that one which should be considered as the LCAO *representation* of the first-order density matrix $\varrho_1(\mathbf{x};\mathbf{x}')$. So for non-orthogonal basis sets, we must distinguish between the concepts of the LCAO expansion and LCAO representation of the first-order density matrix, which are provided by the matrices **P** and **PS**, respectively. (Thus the term "density matrix" often used for the matrix **P**, or its blocks $\mathbf{P}^{\alpha}$ and $\mathbf{P}^{\beta}$, is not very fortunate.)

Some reasons why the matrix **PS** can be most appropriately considered as the proper LCAO representation of the first-order density matrix are as follows.

(1) The matrix **PS** is the matrix describing the linear transformation induced by the density operator $\hat{\varrho}$ associated with the first-order density matrix—c.f. the discussion in (ii) above.

(2) The matrix **PS** is that one which has the correct trace equal to the number of electrons, as it is characteristic for the trace of the first-order density matrix:

$$\mathrm{Tr}(\mathbf{PS}) = \sum_{\mu}(\mathbf{PS})_{\mu\mu} = \mathrm{Tr}\,\varrho_1(\mathbf{x};\mathbf{x}') = \int \varrho_1(\mathbf{x};\mathbf{x})d\mathbf{x} = \int \varrho(\vec{r})dv = N . \tag{4.21}$$

(3) The matrix **PS** is that one which has the correct idempotency property for *single determinant wave functions*:

$$(\mathbf{PS})^2 = \mathbf{PS} . \tag{4.22}$$

This can easily be checked by writing out matrix **P** as a sum of dyadic products $\mathbf{c}^i\mathbf{c}^{i\dagger}$ of the coefficient vectors of some set of orthonormalized molecular orbitals—c.f. Eq. (1.6)—and utilizing their orthonormalization: $\mathbf{c}^{i\dagger}\mathbf{S}\mathbf{c}^j = \delta_{ij}$.

(4) For general (e.g., configuration interaction) wave functions, the eigenvalues and eigenvectors of matrix **PS** give the occupation numbers and LCAO

*Matrix **S** is the overlap matrix of the column vectors forming matrix $\mathbf{C}^{-1}$. As the eigenvectors of matrix **X** are the columns of matrix **C** and they form a whole set of eigenvectors according to the conditions of the theorem, they are linearly independent, which means that the inverse matrix $\mathbf{C}^{-1}$ exists and consists also of linearly independent columns. Therefore their overlap (metric) matrix **S** is positive definite.

coefficients of the *natural spin-orbitals*. In fact, according to the definition of the natural spin-orbitals $\varphi_i(\mathbf{x}) = \sum_\mu c_\mu \chi_\mu(\mathbf{x})$, the first-order density matrix is diagonal in their terms (the ν_i-s are the occupation numbers):

$$\varrho_1(\mathbf{x},\mathbf{x}') = \sum_i \nu_i \varphi_i^*(\mathbf{x}')\varphi_i(\mathbf{x}) , \tag{4.23}$$

which leads to an LCAO expansion (4.9) with the coefficient-matrix

$$\mathbf{P} = \sum_i \nu_i \mathbf{c}^i \mathbf{c}^{i\dagger} . \tag{4.24}$$

Now, the column-vectors $\mathbf{c}^j$ are eigenvectors of the matrix **PS** with the eigenvalues equal to the respective occupation numbers ν_j, owing to the orthonormalization relationship $\mathbf{c}^{i\dagger}\mathbf{S}\mathbf{c}^j = \delta_{ij}$.

Of course, the eigenvalue equation of **PS** is equivalent to the *generalized* eigenvalue equation of $\mathbf{L} = \mathbf{SPS}$; alternatively, the natural spin-orbitals can be found by turning to an auxiliary orthonormalized basis. However, it is important to bear in mind that the solutions of the eigenvalue problem for matrix **P** give the natural spin-orbitals only if the basis is orthonormalized.

The above discussion indicates that matrix **PS** should have a central importance in the LCAO theory. Its important property is that its matrix elements $(\mathbf{PS})_{\mu\nu}$ transform once covariantly (ν) and once contravariantly (μ) with the transformation of the basis orbitals,* guaranteeing the invariance of the trace (4.21). For that reason the Giambiagi group usually applies the tensorial notation Π^μ_ν for them (e.g., [32]).

4.2 Mulliken charges and overlap populations: Invariance

In Sections 3.3 and 3.4 we have obtained Mulliken's atomic and overlap populations by applying the formal "atomic resolution of identity" to the integral of the electron density or by integrating the "atomic electron densities" and "overlap densities," and stressed their special importance as far as the internal mathematical structure of the LCAO theory is concerned. This importance can again be seen by considering the properties of the matrix **PS** giving the LCAO representation of the first-order density matrix $\varrho_1(\mathbf{x};\mathbf{x}')$: according to Eq. (4.21), the total number of electrons is the sum of the diagonal elements

*We call covariant a quantity (vector, matrix) which transforms with the same matrix as the LCAO basis orbitals do (e.g., the overlap matrix is covariant in both its indices), while a contravariant one transforms with the inverse of that matrix (e.g., the orbital coefficients).

$(\mathbf{PS})_{\mu\mu}$. This means that the given spin-orbital $\chi_\mu(\vec{r},s)$ contributes $(\mathbf{PS})_{\mu\mu}$ to the total electron population of the system. Thus the *spatial* orbital $\chi_\mu(\vec{r})$ has a contribution $(\mathbf{P}^\alpha\mathbf{S})_{\mu\mu}+(\mathbf{P}^\beta\mathbf{S})_{\mu\mu}=(\mathbf{DS})_{\mu\mu}$, which is nothing more than Mulliken's *gross orbital population* corresponding to that orbital. Assuming that each basis orbital is assigned to one of the atoms, as is usual in practical calculations, we may sum these quantities belonging to the individual atoms, and get again Mulliken's gross atomic populations

$$Q_A=\sum_{\mu\in A}(\mathbf{DS})_{\mu\mu}\,, \tag{4.25}$$

in agreement with (3.25). Writing out explicitly the summation involved in the matrix product, the terms can be grouped by the second index and one gets the net atomic populations q_{AA} and overlap populations q_{AB}:

$$q_{AA}=\sum_{\mu,\nu\in A}D_{\mu\nu}S_{\nu\mu}\,; \qquad q_{AB}=\sum_{\mu\in A}\sum_{\nu\in B}D_{\mu\nu}S_{\nu\mu}\,, \tag{4.26}$$

in agreement with (3.26). Similarly one obtains the respective atomic and overlap densities shown in Eqs. (3.28), (3.29) by writing out the integrands of the overlap matrices.

As already stressed, one may attribute any physical (or chemical) significance only to quantities which remain invariant when the AOs centered on a given atom are subjected to an arbitrary linear transformation. All quantities resulting in Mulliken's population analysis satisfy this requirement. (This is not necessarily true for the so-called Löwdin-populations, as will be discussed in Section 4.3.)

The rotational–hybridizational invariance of the Mulliken's net atomic and overlap populations (and thus the gross atomic populations representing their sums) has been explicitly demonstrated in [5]. Here we are going to show the invariance of the underlying net atomic and overlap spatial densities (3.29) from which follows the invariance both of the overall atomic density (3.28) as well as that of all the integrated quantities q_{AA}, q_{AB}, and Q_A.

We consider only transformations which mix the orbitals centered on the same atom*; note that for any reasonable atom-centered basis set the overall rotation of the molecule induces such a transformation. In other words, we replace the basis spin-orbitals $\{\chi_\mu\}$ centered on each atom A by some new orbitals $\{\chi'_\nu\}$ according to the formula

$$\chi'_\nu=\sum_{\mu\in A}T^A_{\mu\nu}\chi_\mu\,; \qquad \nu\in A\,. \tag{4.27}$$

*This means that the global transformation matrix of the whole basis has a block-diagonal structure [5] corresponding to the individual atoms.

We request the transformation to be non-singular, so the inverse transformation

$$\chi_\mu = \sum_{\nu\in A} \left[(\mathbf{T}^A)^{-1}\right]_{\nu\mu} \chi'_\nu \,; \qquad \mu \in A\,, \tag{4.28}$$

also exists. The natural spin-orbitals can also be expressed in terms of the new basis orbitals:

$$\varphi_i = \sum_\mu c_\mu \chi_\mu = \sum_A \sum_{\mu\in A} c_\mu \chi_\mu = \sum_A \sum_{\mu\in A} c_\mu \sum_{\nu\in A} \left[(\mathbf{T}^A)^{-1}\right]_{\nu\mu} \chi'_\nu = \sum_A \sum_{\nu\in A} c'_\nu \chi'_\nu\,, \tag{4.29}$$

where we have introduced the grouping of the orbitals according to the atoms to which they belong. Eq. (4.29) indicates that the orbital coefficients in the primed system are

$$c'_\nu = \sum_{\mu\in A} \left[(\mathbf{T}^A)^{-1}\right]_{\nu\mu} c_\mu\,. \tag{4.30}$$

According to these considerations, the elements of the AB block (A=B or A≠B) of the P-matrix (4.24) become in the primed system

$$P^{AB\prime}_{\nu\rho} = \sum_i \nu_i \left(\mathbf{c}^{i\prime}\mathbf{c}^{i\prime\dagger}\right)_{\nu\rho} = \sum_i \nu_i \sum_{\mu\in A}\sum_{\tau\in B} \left[(\mathbf{T}^A)^{-1}\right]_{\nu\mu} c^i_\mu c^{i*}_\tau \left[(\mathbf{T}^{B\dagger})^{-1}\right]_{\tau\rho}\,, \tag{4.31}$$

i.e.,

$$P^{AB\prime}_{\nu\rho} = \sum_{\mu\in A}\sum_{\tau\in B} \left[(\mathbf{T}^A)^{-1}\right]_{\nu\mu} P^{AB}_{\mu\tau} \left[(\mathbf{T}^{B\dagger})^{-1}\right]_{\tau\rho}\,. \tag{4.32}$$

Using Eq. (4.32) and substituting the expression (4.27) for the orbitals, we have for the net atomic density (A=B) and the overlap density (A≠B) in the transformed system, depending on both spatial and spin variables (c.f. Eq. (4.9):

$$\begin{aligned} \varrho^{AB\prime}(\mathbf{x}) &= \varrho_1^{AB\prime}(\mathbf{x};\mathbf{x}) = \sum_{\nu\in A}\sum_{\rho\in B} P^{AB\prime}_{\nu\rho} \chi^{*\prime}_\rho(\mathbf{x}) \chi'_\nu(\mathbf{x}) \\ &= \sum_{\mu,\nu,\eta\in A}\ \sum_{\rho,\tau,\varepsilon\in B} \left[(\mathbf{T}^A)^{-1}\right]_{\nu\mu} P^{AB}_{\mu\tau} \left[(\mathbf{T}^{B\dagger})^{-1}\right]_{\tau\rho} T^A_{\eta\nu} (\mathbf{T}^{B\dagger})_{\rho\varepsilon} \chi^*_\varepsilon(\mathbf{x}) \chi_\eta(\mathbf{x})\,. \end{aligned} \tag{4.33}$$

By performing the summations, all terms containing the transformation matrices $\mathbf{T}^A$, $\mathbf{T}^B$ cancel and one gets simply:

$$\varrho^{AB\prime}(\mathbf{x}) = \sum_{\eta\in A}\sum_{\varepsilon\in B} P^{AB}_{\eta\varepsilon} \chi^*_\varepsilon(\mathbf{x}) \chi_\eta(\mathbf{x}) = \varrho^{AB}(\mathbf{x})\,, \tag{4.34}$$

showing the invariance of the net ($A = B$) and overlap ($A \neq B$) densities and, therefore, that of the respective integrated quantities.

The theoretical importance of the Mulliken's gross overlap population is also shown by the fact [24,26] that it is the expectation value of the *operator of atomic population* $\hat{N}_A$

$$Q_A = \langle \hat{N}_A \rangle \,, \tag{4.35}$$

which can be written down in the "mixed second quantized" framework as

$$\hat{N}_A = \sum_{\mu \in A} \hat{\chi}_\mu^+ \hat{\varphi}_\mu^- \,, \tag{4.36}$$

where $\hat{\chi}_\mu^+$ and $\hat{\varphi}_\mu^-$ are the creation and "effective" annihilation operators corresponding to the non-orthogonal basis orbitals* (Appendix A.1).

When considering the pair of atoms A and B, then there are, in fact, two quantities of the overlap population type; quantity q_{AB} discussed above, and a quite similar one, q_{BA}, which one obtains when interchanging the roles of A and B in the derivations. If one uses real basis functions and real orbital coefficients (which is usually the case, but by far not mandatory), then these two quantities are equal and one often calls overlap population their sum $q_{AB} + q_{BA}$. It is sometimes claimed that the Mulliken's gross atomic populations are obtained by an "arbitrary halving" of this overall diatomic overlap population between the respective atoms. That is, however, not the case; quantities q_{AB} and q_{BA} differ in their origin, they need not be equal in the general, complex case† (then they are complex conjugates) and contribute to the gross atomic populations of atoms A and B, respectively, not because of some arbitrary halving but according to detailed derivations reflecting the internal mathematical structure of the LCAO formalism applied. Similar considerations apply also to the individual orbital populations, of course. That is the case even if one has to keep in mind that the Mulliken's gross orbital population $q_\mu = (\mathbf{DS})_{\mu\mu}$ of an orbital is not strictly confined to the interval $0 \leq q_\mu \leq 2$, which would be the case in an orthonormalized basis. This is

*The operator $\hat{N}_A$ can be considered the second-quantized analogue of the operator-sum $\hat{Q}_A = \sum_{i=1}^N \hat{q}_A(i)$, where $\hat{q}_A(i)$ is operator (3.8) written down for the i-th electron. The expectation value of $\hat{Q}_A$ also equals Mulliken's gross atomic population, in accord with Eq. (3.20). The sum $\sum_A \hat{N}_A$ is the operator of the number of electrons; any N-electron function is its eigenfunction with the eigenvalue N. This is in line with the fact that each $\hat{q}_A(i)$ sums to a unit operator and $\sum_A \hat{Q}_A$ will have N such unit operators. At the same time there seems to be no trivial second quantized analogue of the operator $\sum_{i=1}^N \hat{q}_B^\dagger(i)\hat{q}_A(i)$, the expectation value of which gives, according to Eq. (3.22), the overlap population q_{AB}; this is the case because the product $\hat{N}_B^\dagger \hat{N}_A$ would also have terms which formally correspond to a two-electron operator. (A related point caused some confusion in the literature, too; see the discussion in [33] for details.)

†There are actual problems, e.g., the linear molecules, which are difficult to treat quite correctly by using only real functions and coefficients.

an inherent consequence of the orbital non-orthogonality, and indicates that the mathematically privileged character of the Mulliken populations does not necessarily mean a similar practical adequacy in treating some complex situations.

Mulliken's overlap population is an important quantity, measuring the accumulation of the electronic charge in the bonding region within the framework of the Hilbert-space analysis. This is a very important aspect of the bond formation, which is even measured experimentally in the detailed X-ray crystallographic studies. For appropriate basis sets the Mulliken's overlap populations often give a reasonable measure of the strengths of the chemical bonds. (Of course, only similar bonds may be compared.) From the formal point of view, it is important to stress that Mulliken's overlap population is the simplest quantity that is linear in the interatomic D-matrix elements and has the correct rotational–hybridizational invariance.

4.3 Löwdin charges and the problem of rotational invariance

One often turns to a Löwdin-orthogonalized basis set permitting to avoid the complications caused by the overlap and to keep at the same time the connection with the original (overlapping) AO basis: the orthonormalized functions obtained by using Löwdin's "symmetric orthogonalization" [34] are those which are the closest in least-square sense to the original (overlapping) basis [5,35]. (Introducing an auxiliary Löwdin-orthogonalized basis set is often a useful technique in derivations, too.) Löwdin populations are often very useful interpretative tools. Thus they may be attributed a special importance in the case of studying negative ions: the basis sets used in their calculations necessarily contain diffuse functions lacking any definite atomic character, due to which the Mulliken's populations can appear extremely sensitive to minor changes in the basis set and/or geometry and can exhibit completely unphysical numerical values. In these cases the advantages originating from the orthogonality of the Löwdin basis outweigh the drawback that the individual basis orbitals are not fully localized on the respective atoms but spread to some extent over the whole molecule. It appears, however, that the use of Löwdin populations may meet fundamental difficulties, too.

As is known, Löwdin orthogonalization is performed with the aid of the Hermitian "inverse square root" $\mathbf{S}^{-1/2}$ of the overlap matrix by introducing the new basis orbitals as

$$\chi_\nu^\lambda = \sum_\mu S_{\mu\nu}^{-1/2} \chi_\mu , \tag{4.37}$$

where we have again used the shorthand notation already introduced in Chapter 2. Here and further on the superscript "λ" indicates the quantities pertinent to the Löwdin basis. Owing to the fact that each Löwdin orbital can be considered an orthonormalized counterpart of one of the original AOs, there is a one-to-one correspondence between the two orbital sets, and each Löwdin orbital can also uniquely be assigned to one of the atoms.

The atomic projection operator of type (2.1) in the Löwdin basis is

$$\hat{P}_A^\lambda = \sum_{\mu \in A} |\chi_\mu^\lambda\rangle\langle\chi_\mu^\lambda| , \tag{4.38}$$

and it can play the role of the atomic operator $\hat{\rho}_A$ discussed above, because one can build up a resolution of identity as

$$\sum_A \hat{P}_A^\lambda = \hat{I} . \tag{4.39}$$

As the Löwdin basis is orthonormalized, the atomic population in its terms becomes simply

$$Q_A^\lambda = \sum_{\mu \in A} D_{\mu\mu}^\lambda . \tag{4.40}$$

Similarly to that described in the previous section, the orbital coefficients transform contravariantly with the transformation (4.37) of the basis orbitals (i.e., by matrix $\mathbf{S}^{1/2} = \mathbf{S}^{1/2\dagger}$), so the density matrix $\mathbf{D}$ transforms as

$$\mathbf{D}^\lambda = \mathbf{S}^{1/2}\mathbf{D}\mathbf{S}^{1/2} . \tag{4.41}$$

It has been recognized only recently [36] that Löwdin populations do not, in general, obey rotational-hybridizational invariance; moreover, they are not even rotationally invariant in some cases, and equivalent atoms may be assigned different Löwdin populations. Actually the Löwdin populations are invariant only under those changes which induce *unitary* transformations of the orbitals centered on the individual atoms. This is the case when the AOs which transform between themselves under rotations are orthogonal*; however, this condition is not fulfilled, for instance, if the basis contains the 6 Cartesian *d*-orbitals, as is usual for the popular 6-31G* and 6-31G** basis

*The transformation between two orthogonal basis sets spanning the same subspace is unitary.

sets. No rotational invariance problem occurs for the basis sets using the 5 "pure" d-orbitals. (Similar is the situation with the 7 "pure" vs. 10 Cartesian f-orbitals, etc.)

If one subjects the starting non-orthogonal basis orbitals to a transformation of type (4.27), then the overall overlap matrix $\mathbf{S}$ transforms to

$$\mathbf{S}' = \mathbf{T}^{\dagger}\mathbf{S}\mathbf{T}\,, \tag{4.42}$$

where $\mathbf{T}$ is a block-diagonal matrix containing the blocks $\mathbf{T}^A$ corresponding to the individual atoms along its diagonal. Matrix $\mathbf{S}^{1/2}$ can be defined by the series expansion $\mathbf{S}^{1/2} = \mathbf{1} + \frac{1}{2}\mathbf{s} - \frac{1}{8}\mathbf{s}^2 \ldots$ in the powers of the off-diagonal part of the overlap matrix $\mathbf{s} = \mathbf{S} - \mathbf{1}$. If $\mathbf{T}$ is unitary, matrix $\mathbf{s}$ transforms similarly to (4.42):

$$\mathbf{s}' = \mathbf{S}' - \mathbf{1} = \mathbf{T}^{\dagger}\mathbf{S}\mathbf{T} - \mathbf{1} = \mathbf{T}^{\dagger}(\mathbf{1}+\mathbf{s})\mathbf{T} - \mathbf{1} = \mathbf{T}^{\dagger}\mathbf{T} + \mathbf{T}^{\dagger}\mathbf{s}\mathbf{T} - \mathbf{1} = \mathbf{T}^{\dagger}\mathbf{s}\mathbf{T}\,, \tag{4.43}$$

and the same holds for its powers: if $\mathbf{T}$ is unitary, we have from Eq. (4.43) $(\mathbf{s}')^2 = \mathbf{T}^{\dagger}\mathbf{s}\mathbf{T}\mathbf{T}^{\dagger}\mathbf{s}\mathbf{T} = \mathbf{T}^{\dagger}\mathbf{s}^2\mathbf{T}$ and so on. Thus we obtain that matrix $\mathbf{S}^{1/2}$ transforms in a similar manner, i.e., as

$$(\mathbf{S}')^{1/2} = \mathbf{T}^{\dagger}\mathbf{S}^{1/2}\mathbf{T}\,, \tag{4.44}$$

if, and only if, $\mathbf{T}$ is unitary. In fact, if $\mathbf{T}$ is not unitary, $\mathbf{T}^{\dagger}\mathbf{T} \neq \mathbf{1}$ and the last equality on the right-hand side of Eq. (4.43) does not hold. As $\mathbf{T}$ is block-diagonal, it is unitary if each block $\mathbf{T}^A$ is such.

Alternatively, one can argue that Eq. (4.42) is a similarity transformation conserving the eigenvalues of $\mathbf{S}$ only in the case when $\mathbf{T}$ is unitary, $\mathbf{T}^{\dagger} = \mathbf{T}^{-1}$; a non-unitary $\mathbf{T}$ would change the eigenvalues. As matrix $\mathbf{S}^{1/2}$ is formed by using the eigenvalues of $\mathbf{S}$, one obtains the equality (4.44) by the standard technique of calculating matrix $\mathbf{S}^{1/2}$ *via* the diagonalization of matrix $\mathbf{S}$ if, and only if, $\mathbf{T}$ is unitary. (For a detailed derivation of this type see [36]).

Matrix $\mathbf{D}$ again transforms contravariantly, i.e.,

$$\mathbf{D}' = \mathbf{T}^{-1}\mathbf{D}(\mathbf{T}^{\dagger})^{-1} = \mathbf{T}^{\dagger}\mathbf{D}\mathbf{T}\,, \tag{4.45}$$

if matrix $\mathbf{T}$ is unitary. In that case we get that the Löwdin population of atom A in the transformed (rotated) system remains invariant:

$$\begin{aligned} Q_A^{\lambda\,\prime} &= \sum_{\mu\in A} D_{\mu\mu}^{\lambda\,\prime} = \sum_{\mu\in A} [(\mathbf{S}')^{1/2}\mathbf{D}'(\mathbf{S}')^{1/2}]_{\mu\mu} \\ &= \sum_{\mu\in A} [\mathbf{T}^{\dagger}\mathbf{S}^{1/2}\mathbf{T}\mathbf{T}^{\dagger}\mathbf{D}\mathbf{T}\mathbf{T}^{\dagger}\mathbf{S}^{1/2}\mathbf{T}]_{\mu\mu} = \sum_{\mu\in A} [\mathbf{T}^{\dagger}\mathbf{D}^{\lambda}\mathbf{T}]_{\mu\mu} \\ &= \sum_{\mu\in A}\sum_{\rho,\tau} (\mathbf{T}^{\dagger})_{\mu\rho} D_{\rho\tau}^{\lambda} T_{\tau\mu} = \sum_{\mu,\rho,\tau\in A} T_{\tau\mu} (\mathbf{T}^{\dagger})_{\mu\rho} D_{\rho\tau}^{\lambda} = \sum_{\rho\in A} D_{\rho\rho}^{\lambda} = Q_A^{\lambda}\,, \end{aligned} \tag{4.46}$$

where we have utilized the unitary and block-diagonal nature of **T**, in particular the fact that if $\mu \in A$ then $(\mathbf{T}^{\dagger})_{\mu\rho}$ and $T_{\tau\mu}$ differ from zero only if $\rho \in A$ and $\tau \in A$, respectively, and that the block $\mathbf{T}^{A}$ is unitary, too.

Table 4.1
Mulliken and Löwdin atomic populations of the water molecule[a] calculated by using 6-31G** basis set

Atom	Population				
	Mulliken	Löwdin symmetric arrangement[b]	Löwdin non-symmetric arrangement[c]	Davidson–Löwdin	Improved Löwdin
O	8.67072	8.44351	8.43688	8.19247	8.4106
H_1	0.66464	0.77824	0.78197	0.90377	0.7947
H_2	0.66464	0.77824	0.78114	0.90377	0.7947

[a] R(OH)=0.9437 Å, <HOH = 105.84°
[b] Molecule in the *xy*-plane, the bisector of the HOH angle directed along the axis *x*.
[c] Molecule in the *xy*-plane, one of the OH bonds directed along the axis *x*.

If, however, matrix **T** is not unitary—as is the case when the basis is built up of the 6 Cartesian *d*-orbitals (10 Cartesian *f* orbitals, etc.)—then the usual Löwdin populations become orientation-dependent, indicating that they are lacking any physical meaning and do not represent appropriate tools of analysis. Obviously, the use of a quantity lacking correct rotational invariance cannot be admitted, even if the effect may be not too large—as is the case in the example shown in Table 4.1 In the water molecule the *d*-orbitals are only weakly populated polarization functions, and therefore the discrepancies are minor; however, in the transition metal and uranium compounds studied in [37] they reached some tenths of an electron charge, indicating that the effect can be quite significant.

Invariant versions of Löwdin populations

The remedy in this situation may be the use of the modified Löwdin orthogonalization which was introduced by Davidson [38]. In this scheme one pre-orthogonalizes the orbitals on each atom separately, and performs the Löwdin orthogonalization of the overall basis only subsequently. This leads to rotationally invariant results; also, it is immaterial what orthogonalization algorithm is used to get orthonormalized orbitals on the individual atoms.

This procedure leads to significantly different numbers than does the conventional Löwdin orthogonalization (c.f. Table 4.1).

Most recently we have proposed [39] to perform the orthogonalization of only those basis functions (the sets of 6 d-functions, those of 10 f-functions, etc.) that transform between themselves under rotations of the molecule as a whole. This permits to obtain invariant populations which are numerically rather close to the range in which the usual Löwdin-populations oscillate. (In Table 4.1 the respective numbers are in the column "Improved Löwdin .")

5

Effective AOs and effective minimal basis sets

5.1 Definition of effective AOs

Our qualitative understanding of molecular structure strongly relies on the notion of the minimal basis set. We think in terms that the atoms enter the molecules with their 1*s*, 2*s*, 2*p*, etc. orbitals (or their hybrids), while we do our calculations by using larger and larger basis sets. To resolve this contradiction, we should find proper connections between the simple qualitative picture and the results obtained in the large-scale *ab initio* calculations using extended basis sets. There are some very simple cases in which the Hartree–Fock (or DFT) wave function of a molecule can trivially be described in terms of some *distorted* atomic minimal basis, provided that atom-centered basis sets were used in the calculations. (The term "distorted" is understood with respect to the free atoms' orbitals.) Thus, in the H_2 molecule there is only a single MO and it consists of basis functions belonging to either of the atoms. Therefore, one half of that MO is in the subspace defined by the basis set of one hydrogen atom, and another half is in that of the other hydrogen. That means that there is only a single *effective* AO on each atom, irrespective of how many basis functions were used—the structure of the basis influences only the detailed *form* of that effective AO. Similarly, one has only five MOs in the 10-electron water or methane molecules, so their "truncations" on the oxygen and carbon atom, respectively, define five effective atomic orbitals, i.e., as many as the number of functions in a classical minimal basis. This property ceases to exist in a strict sense when one turns to larger systems (already in the water or methane molecules one gets up to five effective AOs also on the hydrogens) but one can find a good connection with the concept of the minimal basis also in a general case. As will be shown below, one can recognize as many significantly populated *effective* AOs on each atom of ordinary compounds, as many orbitals are in the respective minimal basis set, even if large basis sets are used in the calculations [40–42].

The most adequate method of such an analysis is based on the study of the net atomic electron densities (3.29) and consists of calculating the "natural hybrids" of the individual atoms, first defined by McWeeny [43] for orthogonal basis sets.* An advantage of this approach is that it can also be used without modification for electron-correlated problems.

Similar to the (spatial) natural orbitals which diagonalize the spinless first-order density matrix, the "natural hybrid orbitals" $\{\psi_i^A\}$ of atom A diagonalize the net atomic electron density (3.29):

$$\varrho_{AA}(\vec{r}) = \sum_{\mu,\nu\in A} D_{\mu\nu}\chi_\nu^*(\vec{r})\chi_\mu(\vec{r}) = \sum_{i\in A} \nu_i^A \psi_i^{A*}(\vec{r})\psi_i^A(\vec{r}) \,. \tag{5.1}$$

There is an interesting difference between the present, overlapping case and that of the original scheme of McWeeny [43]: one should not simply diagonalize the intraatomic block of the matrix **D** because that would lead to some spurious non-orthogonal atomic hybrids. Instead, in line with the discussion given in Section 4.1, according to which in the overlapping case the natural orbitals are the eigenvectors of the matrix **PS**, one should solve the non-symmetric eigenvalue problem

$$\mathbf{D}^{AA}\mathbf{S}^{AA}\mathbf{C}^A = \mathbf{C}^A\mathbf{\Lambda}^A \,, \tag{5.2}$$

where $\mathbf{D}^{AA}$ and $\mathbf{S}^{AA}$ are the atomic blocks of the density and overlap matrices, respectively, $\mathbf{\Lambda}^A$ is the diagonal matrix of the atomic natural occupation numbers ν_i^A, and the columns of matrix $\mathbf{C}^A$ contain the expansion coefficients of the atomic natural hybrids in the original basis. Alternatively, one can turn to an auxiliary orthogonal basis on each atom (e.g., by using Löwdin orthogonalization), transform the density matrix to this basis, diagonalize the transformed atomic density matrix, and transform the eigenvectors back to the original basis. This algorithm is accomplished in our free program [47].

The equivalence of the two schemes can easily be checked by multiplying both sides of Eq. (5.2) by $(\mathbf{S}^{AA})^{1/2}$, writing $\mathbf{S}^{AA}$ as the square of $(\mathbf{S}^{AA})^{1/2}$ and denoting $(\mathbf{S}^{AA})^{1/2}\mathbf{C}^A = \mathbf{U}$. Thus, to do the calculations, one should determine for each atom the Hermitian matrices $(\mathbf{S}^{AA})^{1/2}$ and $(\mathbf{S}^{AA})^{-1/2}$; transform the density matrix to the Löwdin basis as $\mathbf{D}^\lambda = (\mathbf{S}^{AA})^{1/2}\mathbf{D}^{AA}(\mathbf{S}^{AA})^{1/2}$; diagonalize the Hermitian matrix $\mathbf{D}^\lambda$ by the unitary matrix $\mathbf{U}$ as $\mathbf{U}^\dagger\mathbf{D}^\lambda\mathbf{U} = \mathbf{\Lambda}^A = diag\{\nu_1^A, \nu_2^A, \ldots\}$; back-transform as $\mathbf{C}^A = (\mathbf{S}^{AA})^{-1/2}\mathbf{U}$, and then one obtains the atomic natural hybrid orbitals in the original basis as $\psi_i^A = \sum_{\mu\in A} C_{\mu i}^A \chi_\mu$.

*The atomic natural hybrids discussed in this section are only in a quite loose connection with the concepts of "natural hybrid orbitals" and "natural atomic orbitals" of Weinhold and coworkers [44–46]; for a detailed discussion of their relationships we refer to [42].

Thus one arrives at a set of *orthonormalized* effective atomic hybrid orbitals ψ_i^A; the eigenvalues obtained in the diagonalization are the respective occupation numbers v_i^A entering the expansion (5.1). As is easy to see, the sum of eigenvalues equals Mulliken's net population of the atom in question.

If one deals with a single determinant wave function, then it is natural to ask what is the relation between the atomic natural hybrid orbitals obtained in the above procedure and the original MOs $\varphi_i(\vec{r})$ of which the wave function of the molecule is built up.* Owing to the invariance of the single determinant wave function with respect to the unitary transformation of the occupied orbitals, the net atomic density $\varrho_{AA}(\vec{r})$ can be written in terms of *any* (canonical or non-canonical) set $\{\varphi_i\}$ of orthogonal MOs as

$$\varrho_{AA}(\vec{r}) = 2\sum_i^{occ.} \varphi_i^{A*}(\vec{r})\varphi_i^A(\vec{r}) , \tag{5.3}$$

where

$$\varphi_i^A = \sum_{\mu\in A} c_\mu^i \chi_\mu , \tag{5.4}$$

is the part of the MO φ_i in which only the intraatomic part is conserved. In the "bra-ket" notations this corresponds to $|\varphi_i^A\rangle = \hat{\rho}_A|\varphi_i\rangle$, where $\hat{\rho}_A$ is the Hilbert-space atomic operator (3.8) of cut-off type.

The atomic "truncations" φ_i^A of orbitals φ_i are, in general, not orthogonal to each other; moreover, they are not even linearly independent if the number of MOs in the molecule exceeds the number of basis orbitals on the given atom. Owing to the unitary freedom of transforming the MOs between themselves, most different atomic "truncations" φ_i^A can be constructed. It appears that there is a unique set of localized orbitals, different for each atom, for which the intraatomic parts φ_i^A of the orbitals are orthogonal to each other—similar to their parent global MOs φ_i—and, up to the normalization, coincide with the atomic natural hybrids ψ_i^A discussed above. In other words, each natural hybrid ψ_i^A with a non-zero occupation number v_i contributes to one, and only one, of these localized orbitals.†

*For the sake of simplicity we consider a closed shell determinant with all occupation numbers equal to 2. Generalization to the UHF case would be straightforward—but then one would have also to look separately for the atomic natural hybrids of spins α and β, respectively, in the previous discussions.

†These localized orbitals are not strictly unique if it happens that there are atomic natural orbitals with degenerate occupation numbers, as that means a unitary freedom of mixing them; in that case, however, one can find the appropriate localized MOs for any choice of the natural orbitals in the degenerate subspace.

For any (canonical or non-canonical) set of MOs $\{\varphi_i\}$ we calculate the norm of their atomic truncation

$$M_i = \langle \varphi_i^A | \varphi_i^A \rangle = \langle \varphi_i | \hat{\rho}_A^\dagger \hat{\rho}_A | \varphi_i \rangle = \sum_{\mu,\nu \in A} c_\mu^{i*} S_{\mu\nu} c_\nu^i . \tag{5.5}$$

One can easily see by integrating Eq. (5.3) that the sum of the quantities M_i is equal to the half of Mulliken's net atomic population of the atom A:

$$q_{AA} = 2 \sum_i^{occ.} M_i . \tag{5.6}$$

(The factor 2 is due to our use of doubly occupied orbitals.) The quantities M_i were first used in the "extrinsic" localization method of Magnasco and Perico [48].

Now, we subject the original (say canonical) *orthonormalized* set of molecular orbitals to a linear transformation

$$\varphi_i' = \sum_j^{occ.} A_{ji} \varphi_j = \sum_j^{occ.} a_j^i \varphi_j , \tag{5.7}$$

requesting the parameter M_i to have a maximal (or, at least, stationary) value for the transformed orbitals. This can be formulated as a standard linear variational problem: one has to find the stationary values of the Rayleigh quotient

$$M_i' = \frac{\langle \varphi_i' | \hat{Q} | \varphi_i' \rangle}{\langle \varphi_i' | \varphi_i' \rangle} = \frac{\mathbf{a}^{i\dagger} \mathbf{Q} \mathbf{a}^i}{\mathbf{a}^{i\dagger} \mathbf{a}^i} , \tag{5.8}$$

where the Hermitian operator $\hat{Q}$ is defined as

$$\hat{Q} = \hat{\rho}_A^\dagger \hat{\rho}_A , \tag{5.9}$$

the elements of the $n_{occ.} \times n_{occ.}$ Hermitian matrix $\mathbf{Q}$ are defined as

$$Q_{ij} = \langle \varphi_i | \hat{Q} | \varphi_j \rangle = \sum_{\mu,\nu \in A} c_\mu^{i*} S_{\mu\nu} c_\nu^j , \tag{5.10}$$

(the c_μ^i-s being the LCAO coefficients of the original orbitals φ_i) and the vector $\mathbf{a}^i$ represents the i-th column of the $n_{occ.} \times n_{occ.}$ transformation matrix $\mathbf{A}$ introduced in Eq. (5.7).

It is well known that the solutions of such a linear variation problem are given by the eigenvectors of the Hermitian eigenvalue equations

$$\mathbf{Q}\mathbf{a}^i = M_i \mathbf{a}^i . \tag{5.11}$$

This gives for the coefficients of the transformed orbitals

$$\varphi_i' = \sum_j^{occ.} a_j^i \varphi_j = \sum_\mu c_\mu^{i\prime} \chi_\mu \ , \tag{5.12}$$

the expansion

$$c_\mu^{i\prime} = \sum_j^{occ.} a_j^i c_\mu^j \ . \tag{5.13}$$

As matrix $\mathbf{Q}$ is Hermitian, its eigenvectors $\mathbf{a}^i$ form columns of a unitary matrix, and consequently the transformed MOs φ_i' are orthonormalized, too. (The original orbitals were assumed orthonormalized.)

Let us now calculate the overlap of atomic "truncations" $\varphi_i^{A\prime}$ of the transformed orbitals φ_i', by taking into account that they are formed with the aid of the coefficients satisfying the eigenvalue equation (5.11):

$$\begin{aligned}\langle \varphi_i^{A\prime} | \varphi_j^{A\prime} \rangle = \langle \varphi_i' | \hat{\rho}_A^\dagger \hat{\rho}_A | \varphi_j' \rangle = \langle \sum_k^{occ.} a_k^i \varphi_k | \hat{Q} | \sum_l^{occ.} a_l^j \varphi_l \rangle = \sum_{k,l}^{occ.} a_k^{i*} Q_{kl} a_l^j \\ = \mathbf{a}^{i\dagger} \mathbf{Q} \mathbf{a}^j = M_j \delta_{ij} \ . \end{aligned} \tag{5.14}$$

This result has several consequences. First of all, it shows that the "truncations" $\varphi_i^{A\prime}$ are orthogonal to each other if at least one of them belongs to a non-zero eigenvalue. Furthermore, the normalization integral $\langle \varphi_i^{A\prime} | \varphi_i^{A\prime} \rangle$ is equal to the eigenvalue M_i, which also means that the orbitals φ_i' belonging to zero eigenvalues do not have any contributions from the orbitals of atom A.*

As Eq. (5.3) can equally well be applied to the transformed set of the orbitals φ_i', too, we may write

$$\varrho_{AA}(\vec{r}) = 2 \sum_i^{occ.} \varphi_i^{A\prime *}(\vec{r}) \varphi_i^{A\prime}(\vec{r}) \ , \tag{5.15}$$

or, introducing the normalized counterpart of the "truncations" as†

$$\psi_i^A(\vec{r}) = \frac{1}{\sqrt{M_i}} \varphi_i^{A\prime}(\vec{r}) \ , \tag{5.16}$$

*If the basis orbitals on the atom A are linearly independent, then the intraatomic part of the overlap integral represents a positive definite metric matrix, and a function with zero norm should have all its coefficients vanish, too.

†In the "bra–ket" notations one has $|\psi_i^A\rangle = (M_i)^{-1/2} \hat{\rho}_A |\varphi_i^{A\prime}\rangle$.

we get

$$\varrho_{AA}(\vec{r}) = 2\sum_{i\in A} M_i \psi_i^{A*}(\vec{r})\psi_i^A(\vec{r}) , \tag{5.17}$$

where the notation $i \in A$ indicates that we have to sum only over the orbitals having non-zero components on the atom in question (i.e., over those that correspond to non-zero M_i values).

Comparison of Eq. (5.17) with Eq. (5.1) shows that we have recovered the natural hybrid orbitals: the orbitals ψ_i^A obtained are orthonormalized and diagonalize the net atomic density ϱ_{AA}; the occupation numbers are given by the eigenvalues M_i as $v_i^A = 2M_i$.

The diagonalization of the matrix **Q** is a simple and quite straightforward technique which works very well in practice—all the calculations in papers [40,42] were done with it. However, for large systems, in which the number of occupied orbitals $n_{occ.}$ can also be large, it is convenient if one can avoid solving equations of dimensions $n_{occ.} \times n_{occ.}$, and reduce the problem to solving equations of dimensions equal to the number of the basis orbitals of atom A. Obviously, that becomes imperative if one wishes to study the analogous Wannier-type orbitals in an *extended* (infinite) system [49]. It appears that the determination of the complete MOs φ_i' can be accomplished by solving merely the matrix equation (5.2) permitting to calculate the atomic natural hybrids.

To start with, we should first discuss that every column of the matrix **D** represents the coefficient vector of an orbital completely lying in the occupied subspace. To see this, we first form the vector $\mathbf{f}^\tau = \mathbf{S}^{-1}\mathbf{e}^\tau$, where $\mathbf{e}^\tau$ is the τ-th unit vector, i.e., a vector, all coefficients of which are zero, except the τ-th which is 1. (In other words, the coefficients of vector $\mathbf{e}^\tau$ are $e_\mu^\tau = \delta_{\mu\tau}$.) As matrix **DS** is, up to a constant factor of 2, the matrix of projection onto the occupied subspace, vector $\mathbf{DSf}^\tau$ is a vector completely lying in the occupied subspace. But substituting the definition of $\mathbf{f}^\tau$ we have $\mathbf{DSf}^\tau = \mathbf{DSS}^{-1}\mathbf{e}^\tau = \mathbf{De}^\tau = \mathbf{d}^\tau$, i.e., the τ-th column of matrix **D**, which is, therefore, a vector in the occupied subspace, indeed. (We have assumed the basis to be linearly independent, so matrix $\mathbf{S}^{-1}$ exists.)

Now, we may assume that we have already performed the transformation of the basis orbitals χ_μ belonging to atom A to the basis of atomic natural hybrids ψ_i^A. In this basis (indicated by prime) the atomic block of the matrix $\mathbf{D}'$ is diagonal, so for each ψ_i^A there is a column $\mathbf{d}^{\tau\prime}$ in the $\mathbf{D}'$ matrix, which has only a single non-zero element (at the τ-th diagonal position in terms of the overall molecular basis) within the block corresponding to atom A, equal to v_i^A. Obviously, this column gives (up to a normalization constant) the coefficients of the orbital φ_i' in that particular (primed) basis. Thus, in order to get the expansion coefficients of the (unnormalized) orbital φ_i' we have to transform this column vector back—now as a vector of orbital coefficients—to the original basis.

When one turns to the basis of the natural hybrids on atom A and leaves unchanged the orbitals of the other atoms, then one performs the transformation of the overall basis with a matrix $\mathbf{T}$ that differs from a unit matrix by having matrix $\mathbf{C}^A$—solution of Eq. (5.2)— at the diagonal block corresponding to atom A. Under transformation of the orbitals by matrix $\mathbf{T}$, the density matrix transforms as $\mathbf{D}' = \mathbf{T}^{-1}\mathbf{D}(\mathbf{T}^\dagger)^{-1}$—c.f. Eq. (4.32). Thus the τ-th column of the transformed density matrix in terms of the original density matrix $\mathbf{D}$ (that calculated in the original LCAO basis) and of the transformation matrix $\mathbf{T}$ can be written as

$$\mathbf{d}^{\tau\prime} = \mathbf{D}'\mathbf{e}^\tau = \mathbf{T}^{-1}\mathbf{D}(\mathbf{T}^\dagger)^{-1}\mathbf{e}^\tau \tag{5.18}$$

Now, if the basis is transformed by matrix $\mathbf{T}$, then the orbital coefficients are transformed by matrix $\mathbf{T}^{-1}$; therefore the *back-transformation* of the orbital coefficients can be done by the inverse of matrix $\mathbf{T}^{-1}$, i.e., by matrix $\mathbf{T}$. Thus we get for the coefficient-vector $\mathbf{d}^\tau$ of the (unnormalized) orbital φ_i' in the original basis

$$\mathbf{d}^\tau = \mathbf{T}\mathbf{d}^{\tau\prime} = \mathbf{D}(\mathbf{T}^\dagger)^{-1}\mathbf{e}^\tau \tag{5.19}$$

Expressing in terms of the coefficients and taking into account that $(\mathbf{e}^\tau)_\eta = \delta_{\tau\eta}$, we get for the coefficients of the (unnormalized) orbital φ_i'

$$d_\mu^\tau = \sum_\rho D_{\mu\rho}[(\mathbf{T}^\dagger)^{-1}]_{\rho\tau} \tag{5.20}$$

Because $\tau \in A$, and matrix $\mathbf{T}$, and therefore also $(\mathbf{T}^\dagger)^{-1}$, is a unit matrix outside the diagonal block corresponding to atom A, the summation can be restricted to $\rho \in A$, and we get:

$$d_\mu^\tau = \sum_{\rho\in A} D_{\mu\rho}[(\mathbf{C}^{A\dagger})^{-1}]_{\rho\tau} = \sum_{\rho\in A} R_{\rho\tau}^A d_\mu^\rho \tag{5.21}$$

(As noted above, the intraatomic block of matrix $\mathbf{T}$ is matrix $\mathbf{C}^A$.)

Thus we may conclude that the coefficients of the molecular orbitals φ_i' we are looking for represent linear combinations of those columns of the original matrix $\mathbf{D}$, which correspond to the atom A in question [40,42]. The transformation matrix $\mathbf{R}^A$ is connected with the solution $\mathbf{C}^A$ of Eq. (5.2) as $\mathbf{R}^A = (\mathbf{C}^{A\dagger})^{-1}$. Taking the adjoint of Eq. (5.2) and multiplying it from both sides by $(\mathbf{C}^{A\dagger})^{-1}$ (i.e., by $\mathbf{R}^A$) we get the equation for $\mathbf{R}^A$ as

$$\mathbf{S}^{AA}\mathbf{D}^{AA}\mathbf{R}^A = \mathbf{R}^A\mathbf{\Lambda}^A\,, \tag{5.22}$$

where it was taken into account that matrices $\mathbf{S}^{AA}$ and $\mathbf{D}^{AA}$ are Hermitian. Similar to Eq. (5.2), this is also a non-symmetric eigenvalue problem with real eigenvalues*; as the two matrices are adjoints of each other, the eigenvectors of the two problems form bi-orthogonal sets, leading to $\mathbf{C}^A\mathbf{R}^A = \mathbf{1}$.

This form of the equations was first derived in Ref. 40 by using a quite different approach: we started with the problem of finding the transformation of the MOs providing stationary value for the quantities M_i—c.f. Eqs. (5.7) to (5.11) above—and arrived at Eq. (5.22) after some lengthy and involved derivations; the equivalence with McWeeny's natural hybrid orbitals was obtained only as a conclusion [42]. The present approach—although, of course, fully equivalent—seems more satisfactory.

In Ref. 49 we used Eq. (5.22) not for an individual atom as reference unit but for a whole elementary cell of an infinite periodic polymer and got Wannier-type orbitals, mostly localized on a given cell. Analogously, in the molecular case, too, one can define A not as a single atom but as a functional group of the molecule and get group natural orbitals and the localized MOs corresponding to them (having the group natural orbitals as their intragroup components). Such a generalization does not require any changes in the formalism; one has simply to consider the union of basis sets centered on the atoms in the group instead of the atomic basis.

A few years ago a complex theoretical construct called "domain averaged Fermi hole" (DAFH) was introduced (e.g., [50]), which in the Hartree–Fock case can be seen to be identical to our localized orbitals corresponding to the natural hybrids [51], while in the correlated case have hardly any other chemical meaning than that they give pretty expensive approximations to the natural hybrid orbitals. So, most recently they were replaced by the "pseudo-DAHF-s" which essentially represent 3D (AIM) analogues of the natural hybrids [52].

Before turning to the discussion of some example calculations, it is worth mentioning that the above formalism can be applied not only if the target of optimization is the Mulliken's net population M_i connected with the orbital in question; any bilinear Hermitian localization functional of the MOs can be used [42] instead. The conceptual similarity to the results considered above becomes especially evident if one is able to formulate an operator $\hat{Q} = \hat{\rho}_A^\dagger \hat{\rho}_A$ introduced in Eq. (5.9) with an adequate definition of the atomic operator $\hat{\rho}_A$. In particular, one can obtain effective atomic orbitals in the framework of the 3D analysis; they are discussed in Section 9.4.

*If one uses the algorithm of the solution discussed on page 50, then one simply has $\mathbf{R}^A = (\mathbf{S}^{AA})^{1/2}\mathbf{U}$. (This is utilized in our program [47].)

5.2 Some calculations of effective AOs

We consider it to be of utmost conceptual importance that for every atom of all "conventional" molecules the number of effective AOs having appreciable occupation numbers has always been found equal to the number of orbitals in a formal minimal basis set; all the other orbitals follow only after (at least) an order of magnitude drop in the occupation numbers. That means that one can *extract* an effective minimal basis from the results of large-scale *ab initio* calculations using pretty large basis sets. (The results become somewhat less "well-cut" if the basis contains diffuse functions; as already noted, the latter do not have any pronounced atomic character and are not very adequate for performing Hilbert space analyses.) These results show that the ubiquitous qualitative considerations based on the presence and behavior of different atomic orbitals in a molecule (e.g., changes in their populations under different interactions) are indeed legitimate. Of course, the effective AOs extracted *a posteriori* from the results of calculations are not those of the free atoms, nor even strictly pure *s*, *p*, etc., orbitals; but usually they can be characterized as either having a dominating *s* or *p* character or as *sp* hybrids. Of course, all of them can have some *d* (*f*) admixture. These two cases are well illustrated on Figures 5.1 and 5.2 for methane and ammonia, respectively.

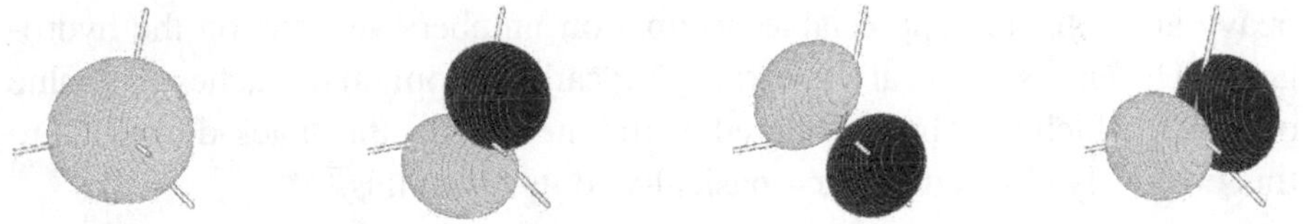

Figure 5.1
The four valence effective AOs of the methane molecule, extracted from a Hartree–Fock calculation using cc-pVTZ basis set.

Figures 5.3, and 5.4 display the natural occupation numbers (in descending order) of the carbon and hydrogen atoms of the benzene molecule, respectively, calculated by using the cc-pVTZ basis set. In this basis each carbon atom carries 30 basis functions (four *s*, three sets of *p*, two sets of *d*, and one set of *f* orbitals), out of which one can construct just 5 sets of effective AOs

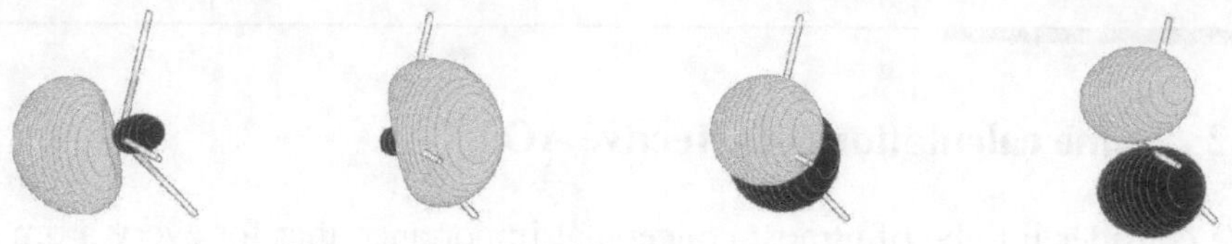

Figure 5.2
The four valence effective AOs of the ammonia molecule, extracted from a Hartree–Fock calculation using cc-pVTZ basis set.

with appreciable occupation numbers: the 6-th AO has the tiny $\nu_6^A = 0.0025$. The first five effective AOs clearly correspond to the one core and four valence orbitals of the elementary textbooks. Analogously, out of the 14 basis function (three *s*, two sets of *p*, and one of *d* orbitals) a single appreciably populated effective AO can be constructed for each hydrogen atom: the second effective AO has only $\nu_2^A = 0.0017$. This sharp drop in the natural occupation numbers is not peculiar for the benzene molecule: in ethane one has, for instance, $\nu_6^A = 0.0015$ and $\nu_2^A = 0.0023$ for the carbon and hydrogen atoms, respectively, if the same cc-pVTZ basis is used—and so on.

Figure 5.5 shows the occupation numbers of the effective orbitals of the different atoms in the glycine molecule* by using a more compact way of presentation. One can see that again there are five effective AOs on each heavy atom, having appreciable occupation numbers and one on the hydrogens. The largest ν_6^A is at the carboxylic carbon atom, and reaches the value of 0.021, which may be considered as the measure of the "back-donation" to this positively charged center—basically to its *d*-orbitals.†

Another characteristic feature, recurring in most different systems is that the lone pair orbitals of oxygen, nitrogen, sulfur, chlorine, etc., atoms often have net occupation numbers ν_i^A exceeding two; this is then compensated by negative overlap populations with the orbitals of other atoms. But that is not necessarily the case: in the glycine molecule this effect is exhibited by the

*We show first the same examples of benzene and glycine as were discussed in [42] but the basis set used here is significantly larger [53]. The conclusions do not differ, however. (Note that in Ref. 42 the quantity M_i was displayed which is half of the value ν_i^A used on the figures here.)

†This carbon atom has a significantly greater resulting *d*-population than does the α-carbon: the aggregated Mulliken's gross orbital population of the d-orbitals is ca. 0.294. This is in agreement in some peculiarities in its valence behavior to be noted in Section 6.12.

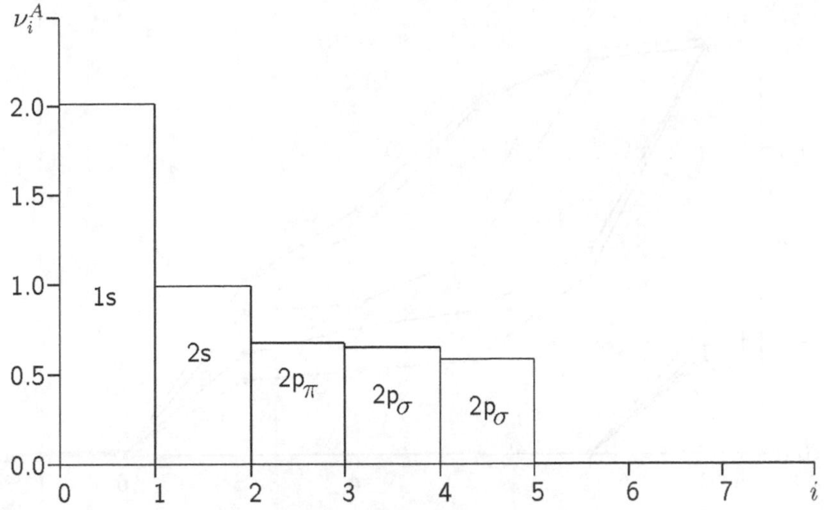

Figure 5.3

Occupation numbers (in descending order) and characteristics of the effective atomic orbitals of carbon atoms in benzene, calculated by using cc-pVTZ basis set.

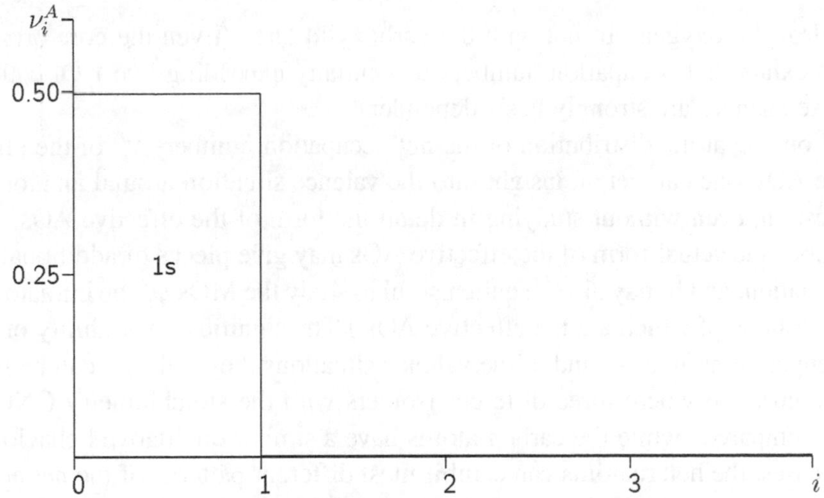

Figure 5.4

Occupation numbers (in descending order) and characteristics of the effective atomic orbitals of hydrogen atoms in benzene, calculated by using cc-pVTZ basis set.

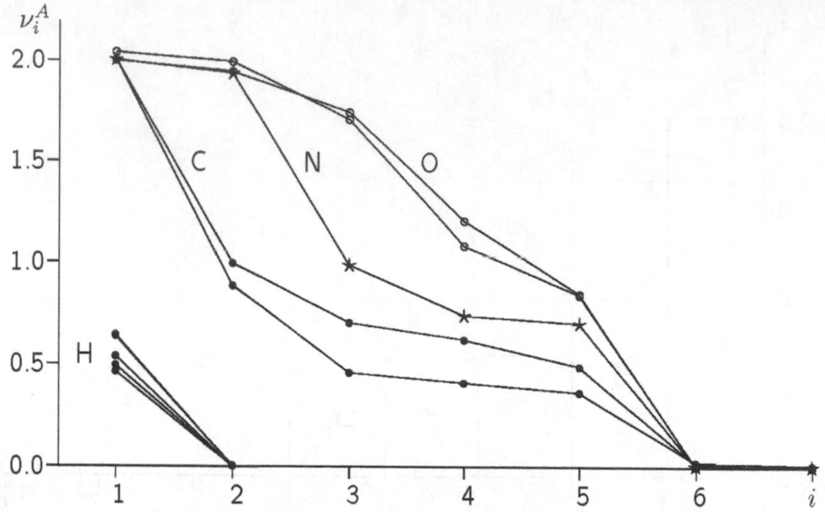

Figure 5.5
Occupation numbers (in descending order) of the effective atomic orbitals of different atoms in glycine molecule, calculated by using cc-pVTZ basis set. From [53] with permission from Wiley.

hydroxylic oxygen but not with the carboxylic one. (Even the core orbitals can exhibit net occupation numbers tangentially exceeding two.) Of course, these nuances are strongly basis-dependent.

Looking at the distribution of the net occupation numbers v_i^A of the effective AOs, one can get an insight into the valence situation around an atom in question, even without studying in detail the form of the effective AOs. (Of course, the actual form of the effective AOs may give pieces of additional information, and it may also be quite useful to study the MOs φ_i' the intraatomic truncations of which are the effective AOs.) The significant variability of the occupation numbers—and of the valence situations they reflect—can be seen in Figure 5.6 where three different isomers with the stoichiometry $CNOH_3$ are compared: while the carbon atoms have a similar qualitative behavior in all cases, the heteroatoms can exhibit most different patterns of the net occupation numbers depending on their environment.

An important conclusion which can be drawn on the basis of Figure 5.6c is that even the nitrone molecule CH_2=NH=O, with its unusual "hypervalent" structure (the nitrogen atom has a formal valence of five) can be described by using the same five effective AOs on each heavy atom: the largest v_6^A is on nitrogen, but it is only 0.0103. Such a behavior may be expected for any

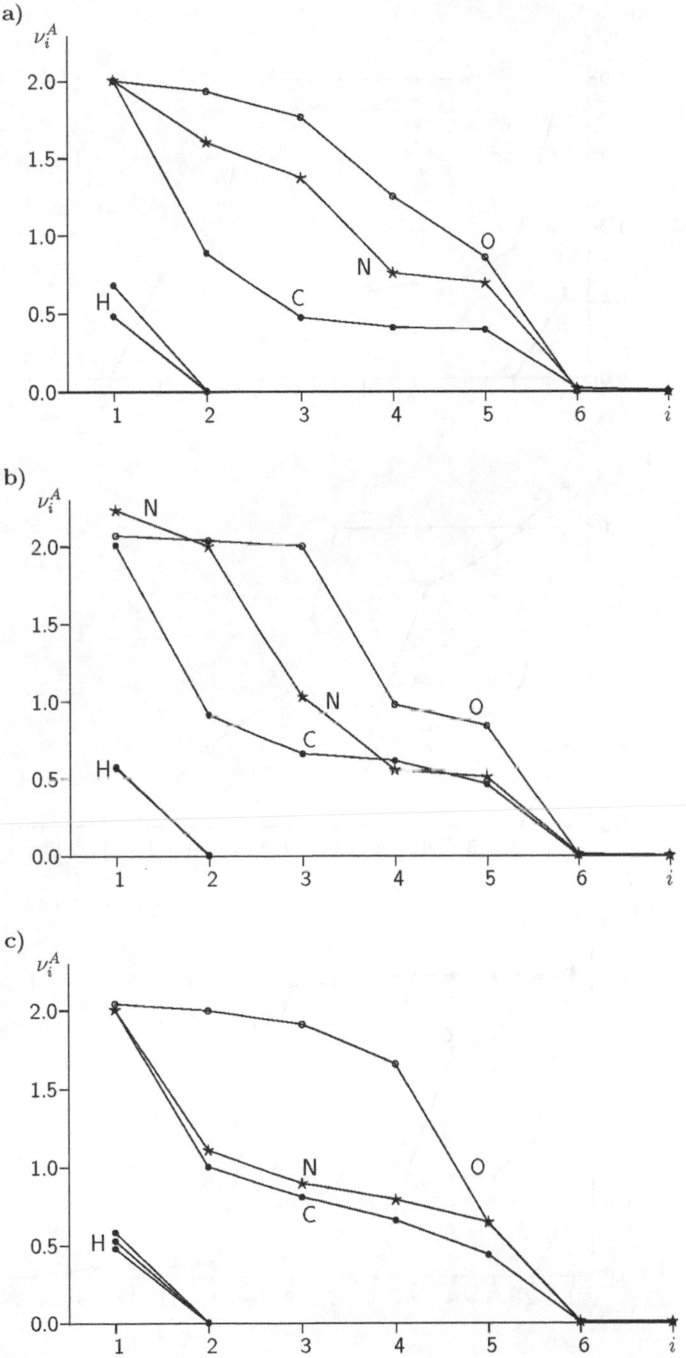

Figure 5.6

Occupation numbers (in descending order) of effective atomic orbitals in a) formamide, b) nitrosomethane, and c) nitrone molecules, calculated by using cc-pVTZ basis set.

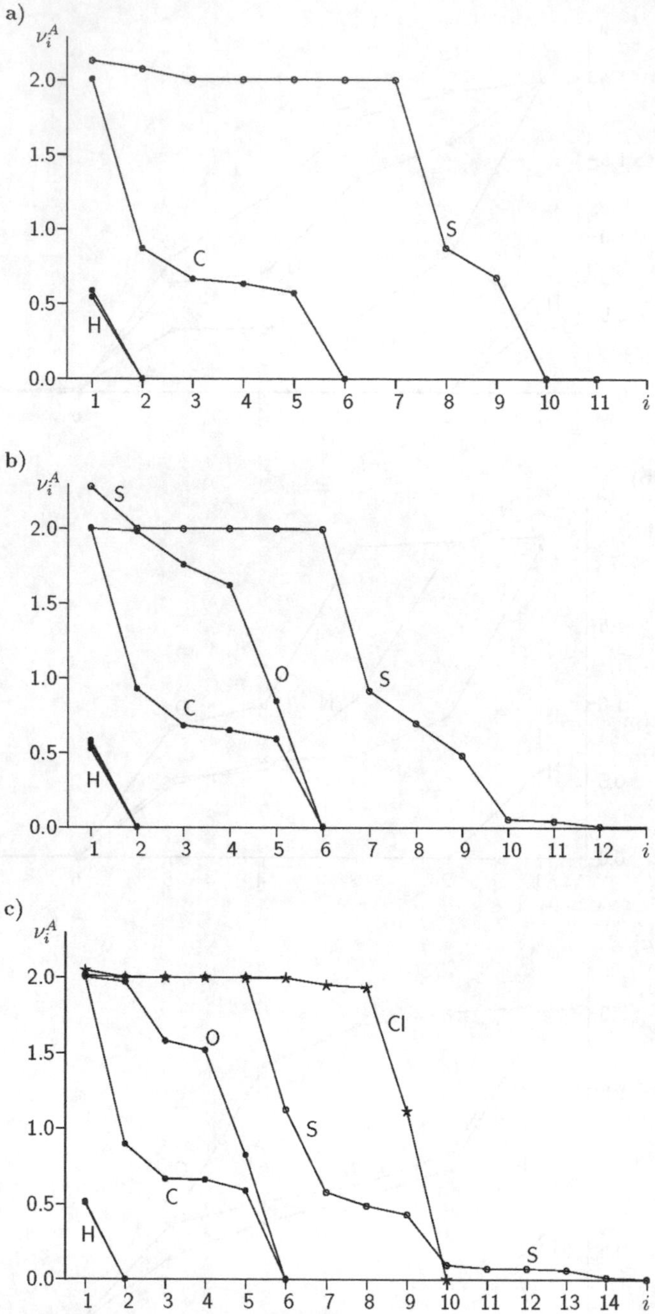

Figure 5.7

Occupation numbers (in descending order) of effective atomic orbitals in a) $(CH_3)_2S$, b) $(CH_3)_2SO$, and c) CH_3SO_2Cl, calculated by using cc-pVTZ basis set. Part c): From [53] with permission from Wiley.

compounds containing only hydrogen and first row atoms Li to F.

The situation, however, changes radically when turning to the second row atoms, having relatively low lying atomic 3*d*-orbitals. Figure 5.7 compares the occupation numbers obtained for dimethyl-sulphide, dimethyl-sulfoxide, and methyl-sulphonyl chloride. As to the dimethyl-sulphide results, we essentially have to mention only that the number of effective AOs of the sulfur is increased by the appearance of the additional core (2*sp*) shell. The sulfur has two bulky lone pairs with significant delocalization to the methyl groups, leading to v_i^A-s exceeding two, one σ, one π; in the oxygen analogue dimethyl-ether there is only one (the π) $v_i^A > 2$, but that is the most pronounced difference.

A qualitatively different behavior is observed for the hypervalent sulfur atoms of formal valence four or six. We see that the net occupation numbers v_i^A do not fall to nearly zero as do the first row elements or divalent sulfur, but there is a significant "shoulder" of two and four occupation numbers for the dimethyl-sulfoxide (DMSO) and methyl-sulphonyl chloride, respectively. (Quite similar "shoulders" have also been observed for other systems with four- and six-valent sulfur, by using different basis sets [40].) The occupation numbers of 0.055 and 0.042 in DMSO and, especially, those between 0.098 and 0.063 for CH_3SO_2Cl are small, but by far not negligible. And these "additional" effective AOs are constituted almost exclusively of *d*-type basis orbitals, similar to the case considered in [40]. These results can be interpreted either as a very significant back-donation to the *d*-orbitals of the strongly positive sulfur or one can attribute some valence character to the sulfur *d*-orbitals in the hypervalent case, but the bonds formed with them are very strongly polarized, leaving only a small portion of the electronic charge on the sulfur. These are, however, only different verbal descriptions of the same situation in which the positive ionic character of the sulfur and the significant role of the *d*-orbitals are equally important. One should keep in mind that the v_i^A values represent net populations. There is also a significant overlap population of the *d*-orbitals in question, significantly increasing the role of these "additional" effective AOs in determining the valence situation around the sulfur. Thus the overall gross *d*-orbital population of sulfur is 0.516 in DMSO and as large as 1.044 in CH_3SO_2Cl—while it is only 0.143 in dimethyl-sulphide, indicating a qualitative difference in the role of *d*-orbitals in these cases.* We

*The value of 0.143 obtained for the *d*-population of divalent sulfur is comparable with the *d* population 0.085 of the carbon atom in the same molecule—for which *d* orbitals clearly play the role of polarization functions only—rather than with the values obtained for the four- and six-valent sulfur. (For comparison: using the same basis, the *d*-populations in dimethyl-ether are 0.037 on the oxygen and 0.115 on the carbon.)

shall return to this point in Section 6.12.

Of course, the above considerations are in full accord with the facts known for a long time (e.g., [54–56]) that the presence of d-orbitals in the basis is imperative to get even qualitatively correct geometries for molecules of hypervalent sulfur (or selenium, tellurium [57]) compounds, while for first row elements and two-valent sulfur they are not nearly as important.

We end this section by briefly mentioning that the effective AOs obtained in the manner discussed above are indeed those which are relevant for the systems studied. One can build up wave functions in which only these basis orbitals are conserved without causing a significant error in the energy. We have tested this in the following manner. All the orbitals were transformed to the basis of effective AOs, and we simply zeroed all the coefficients except those corresponding to the effective minimal basis of every atom. The orbital coefficients were not re-optimized, and the energy has been calculated after re-orthonormalizing the truncated MOs. The energy obtained has been slightly higher than the full SCF energy, of course, but not too much. Then, repeating the procedure by adding the omitted basis orbitals one by one, we got a very smooth convergence to the exact SCF energy in the case of "conventional" molecules. (The convergence could be sped up, of course, by doing a complete SCF calculation for every truncated basis set, i.e., re-optimizing the orbital coefficients after the truncation.) In the hypervalent sulfur compounds, however, the energy had a strong jump if the effective AOs of d-type corresponding to the "shoulders" on the curves were omitted; the smooth behavior started only after they were added to the minimal basis set.

6

Bond order and valence indices in the Hilbert space

6.1 Predecessor: The Wiberg index in CNDO theory

It was realized very early that the use of the Coulson bond order index discussed in Section 1.2 is restricted to the π-electron theories because in the general case one has more than one basis function on each atom. As a consequence, neither the individual two-center D-matrix elements nor their sum for a given pair of atoms obey the required invariance under the rotation of the molecular system with respect of the external laboratory system of coordinates. The quantity proposed for identifying chemical bonds had been Mulliken's overlap population q_{AB}—see the right-hand side of Eq. (3.26) or Eq. (4.26)—which is linear in the interatomic D-matrix elements and has the proper invariance property. Mulliken's overlap populations often are indeed useful measures of the bond strengths. However, the overlap population does not take the values of (about) one, two, or three for the classical single, double, and triple chemical bonds, respectively. Therefore, it cannot be identified with the classical notion of bond order (multiplicity). As will be seen, in order to find a quantum mechanical counterpart of the latter, one needs a quantity which is quadratic in the D-matrix elements.

In 1967, applying the then novel and popular semiempirical CNDO method of Pople and Segal, Wiberg [58] met the difficulty that in that formalism all the basis orbitals are assumed orthonormalized (e.g., they can be considered as originating from a tacit Löwdin orthogonalization), so the overlap population between each pair of atoms vanishes identically. Thus Wiberg had to look for some other quantity characterizing bond strengths. On a heuristic basis, i.e., without any deeper theoretical explanations, Wiberg proposed the use of the "bond index" now bearing his name:

$$W_{AB} = \sum_{\mu \in A} \sum_{\nu \in B} |D_{\mu\nu}|^2 , \tag{6.1}$$

which is necessarily positive. (Wiberg did not consider the possibility of complex orbitals and density matrix elements; in his—verbal only—definition he simply spoke about the sum of squares.)

It had been found that the values of this index calculated for different bonds in "conventional" molecules are close to the classical values of the *multiplicity* of the respective chemical bonds, and even their minor changes can be correlated with the changes in the strengths of the bonds.

Although Wiberg did not investigate the invariance problem, fortunately enough it appears that the Wiberg index (6.1) *is rotationally invariant* [59]. This is the case because two orthogonal basis sets spanning the same subspace are always related to each other by a *unitary* transformation, and the rotation induces such a one.* To see that invariance, we utilize the Hermiticity of the D-matrix and write Eq. (6.1) as

$$W_{AB} = \sum_{\mu\in A}\sum_{\nu\in B} D_{\mu\nu}D_{\nu\mu}\,, \tag{6.2}$$

We need the matrix elements $D'_{\mu\nu}$ obtained after the rotation transformation. This transformation is given by the expression (4.32); for the sake of simplicity we shall omit here the superscripts A, B referring to the sub-blocks of the D-matrix (the restrictions over the summations are sufficient) and introduce the simplified notations $\mathbf{V}^A$ and $\mathbf{V}^B$ for the unitary matrices $(\mathbf{T}^A)^{-1}$ and $(\mathbf{T}^B)^{-1}$, respectively. Then one can prove the invariance by writing

$$\sum_{\mu\in A}\sum_{\nu\in B} D'_{\mu\nu}D'_{\nu\mu} = \sum_{\mu,\rho,\sigma\in A}\ \sum_{\nu,\tau,\eta\in B} V^A_{\mu\rho}D_{\rho\tau}V^{B*}_{\nu\tau}V^B_{\nu\eta}D_{\eta\sigma}V^{A*}_{\mu\sigma} = \sum_{\rho\in A}\sum_{\tau\in B} D_{\rho\tau}D_{\tau\rho}\,, \tag{6.3}$$

where we have utilized a fact that the columns of a unitary matrix are orthonormal: $\sum_{\mu\in A} V^A_{\mu\rho}V^{A*}_{\mu\sigma} = \delta_{\rho\sigma}$ and $\sum_{\nu\in B} V^B_{\nu\eta}V^{B*}_{\nu\tau} = \delta_{\eta\tau}$.

The importance of the Wiberg index is given by a fact which cannot be overemphasized: in the CNDO framework—and, in general, for all theories in which the basis orbitals are orthonormalized—it represents a direct measure of bond multiplicity and gives numbers in very good agreement with the chemists' intuitive concepts. Borisova and Semenov [59] proved that for the singlet states of the first row homonuclear diatomics (except C_2 having a peculiar electronic structure [60]) the Wiberg index equals the "chemist's bond order"

$$B = \frac{N_{bond} - N_{antibond}}{2}\,, \tag{6.4}$$

*The Wiberg index is not invariant under non-unitary hybridization transformations, while the general bond order index Eqs. (6.15), (6.17) is—see page 72.

where N_{bond} and $N_{antibond}$ are the number of electrons occupying bonding and antibonding orbitals,* respectively—i.e., it assumes the ideal integer values 1, 2, or 3.

The idea of the proof can be caught by considering the hybrid AOs of which the individual molecular orbitals are built up. Owing to the symmetry of the homonuclear diatomics, each MO is the (normalized) sum or difference of the hybrids of the two atoms, which are symmetry pairs of each other. Usually there are pairs of both bonding and antibonding MOs formed of the same hybrids, and either both of them are occupied in the SCF wave function, or only the bonding combination is occupied and the antibonding one is empty. All the matrices involved are block-diagonal in terms of these hybrids, which permits to exclude all cross terms between the different orbitals, so their contributions may be considered separately. This means that it is sufficient to discuss explicitly the case when the basis consists of only two orbitals χ_μ^A and χ_μ^B centered on the atoms A and B, respectively, and the resulting Wiberg index is the sum of the values obtained for the individual sub-problems.

If we have the bonding orbital $\varphi_\mu^b = \frac{1}{\sqrt{2}}(\chi_\mu^A + \chi_\mu^B)$ doubly occupied (and the respective antibonding orbital empty) then, as is easy to calculate directly, the relevant sub-block of matrix **D** will be

$$\mathbf{D}^\mu = \begin{pmatrix} 1 & 1 \\ 1 & 1 \end{pmatrix} , \tag{6.5}$$

giving a contribution of unity to the Wiberg index. If, however, the antibonding orbital $\varphi_\mu^a = \frac{1}{\sqrt{2}}(\chi_\mu^A - \chi_\mu^B)$ is also doubly occupied, then we will have

$$\mathbf{D}^\mu = \begin{pmatrix} 2 & 0 \\ 0 & 2 \end{pmatrix} , \tag{6.6}$$

and the hybrids in question do not contribute to the resulting Wiberg index. This last result can simply be explained by utilizing the invariance of determinant wave functions with respect to unitary transformations of the occupied orbitals. In every case when the bonding and the antibonding combinations of some atomic hybrids are both occupied, one can perform the unitary transformation leading to the sum and difference of these MOs. In this manner one obtains a pair of doubly occupied localized orbitals, each of which is

*The fact that one should distinguish between bonding and antibonding orbitals explains the seeming contradiction that usually there are "too many" valence electrons to assume that each pair of electrons occupying a two center molecular orbital in a diatomics corresponds to a chemical bond—although obviously each such MO bears a pair of electrons delocalized between the two atoms, at least in the standard canonical orbital description.

fully concentrated on one atom. These strictly atomic lone pair orbitals (hybrids) do not, of course, contribute to the resulting Wiberg index, and the latter becomes equal to the number of doubly occupied bonding orbitals the antibonding counterparts of which are empty [59]. This result is in full accord with the original Lewis electron pair picture—electrons fully localized at one of the atoms need not be counted when the bonding is considered.*

Wiberg index and exchange

There is a very important connection between the Wiberg index and the *exchange effects*, probably first stressed in [24]. The point is that, owing to the "zero differential overlap" (ZDO) approximation applied, the CNDO model uses only one- and two-center integrals, thus the CNDO energy has a natural breakdown into one- and two-center energy components [61]. For our purpose it is sufficient to consider the two-electron part $E_{(2)}$ of the closed shell Hartree–Fock energy [5]

$$\begin{aligned} E_{(2)} &= \sum_{i,j}^{occ.} \left(2[\varphi_i(1)\varphi_j(2)|\varphi_i(1)\varphi_j(2)] - [\varphi_i(1)\varphi_j(2)|\varphi_j(1)\varphi_i(2)]\right) \\ &= \frac{1}{2} \sum_{\mu,\nu,\rho,\tau} \left(D_{\rho\mu}D_{\tau\nu} - \frac{1}{2}D_{\tau\mu}D_{\rho\nu}\right)[\mu\nu|\rho\tau], \end{aligned} \tag{6.7}$$

where $[\mu\nu|\rho\tau]$ is the shorthand for the integral $[\chi_\mu(1)\chi_\nu(2)|\chi_\rho(1)\chi_\tau(2)]$ over the basis orbitals. The ZDO approximation used in the CNDO framework means that only those integrals are conserved in which the two orbitals corresponding to the given electron i $(i = 1, 2)$ coincide:

$$[\mu\nu|\rho\tau] = \gamma_{AB}\delta_{\mu\rho}\delta_{\nu\tau}\,, \qquad \mu \in A;\ \nu \in B\,, \tag{6.8}$$

γ_{AB} being the interatomic electron-electron repulsion integral of the CNDO scheme.† Thus the two-electron part of the energy (6.7) becomes

$$E_{(2)}^{\text{CNDO}} = \frac{1}{2}\sum_{A,B}\sum_{\mu\in A}\sum_{\nu\in B}\left(D_{\mu\mu}D_{\nu\nu} - \frac{1}{2}D_{\nu\mu}D_{\mu\nu}\right)\gamma_{AB}\,. \tag{6.9}$$

The second term in the parentheses originates from the exchange part of the Hartree–Fock energy. Its inspection shows that for a given pair of atoms A and B the diatomic energy term originating from the exchange part of the

*The above considerations do not apply to C_2 because in its case there is a pair of occupied bonding and antibonding orbitals that are not formed of the same hybrids.

†In the CNDO theory there is only a single such parameter for every pair of atoms of main group elements [18]. Thus γ_{AB} depends only on the nature of the atoms on which orbitals χ_μ, χ_ν are centered and on the interatomic distance.

Hartree–Fock energy is proportional to the respective diatomic Wiberg index [24]:

$$E_{AB,exch} = -\frac{1}{2}\gamma_{AB}W_{AB} , \qquad (6.10)$$

(A factor 2 appears due to the fact that the sums over the atoms A, B in Eq. (6.9) run independently.) This relationship connects the Wiberg index first introduced in a quite *ad hoc* manner to the underlying physics.

6.2 Exchange and bonding

But why and how are bond multiplicity and exchange energy interrelated? To see this, let us first consider the simplest H_2 molecule treated at the CNDO level. In that case there is a single doubly occupied MO $\varphi = \frac{1}{\sqrt{2}}(\chi_a + \chi_b)$ representing a normalized sum of the two atomic orbitals χ_a, χ_b, and the overall two-electron energy, in accord with Eq. (6.9) is simply $\frac{1}{2}(\gamma_{AA} + \gamma_{BB} + \gamma_{AB})$, as matrix **D** looks like that in Eq. (6.5) and the summations over A and B run independently. On the other hand, there are two H-atoms each with electronic charge of one, and the electrostatic two-center Coulomb repulsion between them is γ_{AB}, which is twice as large as the true two-center term of the electron-electron interaction energy. The difference is contained in the exchange term, which in this simplest case reflects the fact that *the electron does not interact with itself.* As the electrons occupy *delocalized* orbitals, each electron contributes to the charges on both atoms A and B. Thus the pure electrostatic interaction between the resulting charges on the two atoms contains terms which originate from the given electron interacting with itself, i.e., with that part of the same electron which is delocalized to the partner atom. Therefore the overall Coulomb repulsion should be reduced—this fact is accounted for by the exchange part of the energy. If, however, both the bonding and antibonding combinations of χ_a, χ_b are occupied, then, as discussed above, we have, in fact, two localized electron pairs on the two atoms (like in the He dimer) and there is no such self-interaction which needs to be corrected. So there is no diatomic exchange and no diatomic bonding. As discussed in the context of Wiberg index, in the simple homonuclear diatomics every pair of hybrids forming a bonding MO, or a pair of bonding and antibonding MOs, can be treated independently and then one can simply sum the results obtained for bond orders. Thus we get that the exchange energy is proportional to the number of bonding electron pairs the antibonding counterparts of which are not occupied—that is to the Wiberg index between the two atoms. In more complex cases the Wiberg index measures some effective number of bonding electron pairs, as determined by the contribution of the different MOs to the diatomic exchange energy in the CNDO model of the molecule.

In this context it is to be stressed that the self-repulsion of electrons can be separated from other effects so nicely only for two electrons (like in H_2), in all other cases one has to make the reservation that only the sum of all exchange terms (comprising self-repulsion and "true" exchange) has the unitary invariance which is required for every physical quantity calculated at the Hartree–Fock level. In the simple CNDO model of homonuclear diatomics based on the proper atomic hybrids, all the diatomic exchange energy may be attributed to self-repulsion when the system is treated as a combination of quasi-independent two-electron systems, as above, but not if the wave function is described by a set of molecular orbitals subjected to an arbitrary unitary transformation.*

Before turning to the discussion of the general (*ab initio*) bond order index, we should mention that Borisova and Semenov [62] have also considered the generalization of the Wiberg index for open-shell systems and proposed the formula

$$W_{AB} = \sum_{\mu \in A} \sum_{\nu \in B} \left(|D_{\mu\nu}|^2 + |P^s_{\mu\nu}|^2 \right), \tag{6.11}$$

containing both the total spinless density matrix $\mathbf{D} = \mathbf{P}^\alpha + \mathbf{P}^\beta$ and the spin density matrix $\mathbf{P}^s = \mathbf{P}^\alpha - \mathbf{P}^\beta$. That is necessary in order to get the proper half-integer values for the Wiberg index in open-shell diatomic systems—e.g., the value of 1/2 for H_2^+.

6.3 Exchange density and the bond order

In Section 3.4 we have discussed that in the single determinant case the exchange density is normalized to the number of electrons—c.f. Eq. (3.44):

$$\iint \varrho_2^x(\vec{r},\vec{r}\,')dvdv' = N. \tag{6.12}$$

*In the first line of Eq. (6.7) the sum $-\sum_i [\varphi_i\varphi_i|\varphi_i\varphi_i]$ present in the energy formula (corresponding to the case $i = j$) accounts for the self-repulsion, but only the overall sum $-\sum_{i,j} [\varphi_i\varphi_j|\varphi_j\varphi_i]$ has unitary invariance. In the CNDO case the terms with $i \neq j$ vanish as a consequence of the ZDO approximation if, and only if, one uses the description by the hybrids discussed.

and the integral at the left-hand side can be decomposed according to the centers involved with the aid of the atomic operators $\hat{\rho}_A$; if Hilbert-space analysis is used, one gets Eq. (3.49), i.e.,

$$\iint \varrho_2^x(\vec{r},\vec{r}')dvdv' = \sum_{A,B}\sum_{\mu\in A}\sum_{\nu\in B}\left[(\mathbf{P}^\alpha\mathbf{S})_{\mu\nu}(\mathbf{P}^\alpha\mathbf{S})_{\nu\mu} + (\mathbf{P}^\beta\mathbf{S})_{\mu\nu}(\mathbf{P}^\beta\mathbf{S})_{\nu\mu}\right] . \tag{6.13}$$

This can trivially be written, interchanging some summation indices, also as

$$\begin{aligned}\iint \varrho_2^x(\vec{r},\vec{r}')dvdv' &= \sum_{A}\sum_{\mu,\nu\in A}\left[(\mathbf{P}^\alpha\mathbf{S})_{\mu\nu}(\mathbf{P}^\alpha\mathbf{S})_{\nu\mu} + (\mathbf{P}^\beta\mathbf{S})_{\mu\nu}(\mathbf{P}^\beta\mathbf{S})_{\nu\mu}\right] \\ &+ 2\sum_{A<B}\sum_{\mu\in A}\sum_{\nu\in B}\left[(\mathbf{P}^\alpha\mathbf{S})_{\mu\nu}(\mathbf{P}^\alpha\mathbf{S})_{\nu\mu} + (\mathbf{P}^\beta\mathbf{S})_{\mu\nu}(\mathbf{P}^\beta\mathbf{S})_{\nu\mu}\right] .\end{aligned} \tag{6.14}$$

The two-center contribution to this sum has been *defined* [63, 64] as the bond order B_{AB} between the respective atoms A and B:

$$B_{AB} = 2\sum_{\mu\in A}\sum_{\nu\in B}\left[(\mathbf{P}^\alpha\mathbf{S})_{\mu\nu}(\mathbf{P}^\alpha\mathbf{S})_{\nu\mu} + (\mathbf{P}^\beta\mathbf{S})_{\mu\nu}(\mathbf{P}^\beta\mathbf{S})_{\nu\mu}\right] . \tag{6.15}$$

In terms of the spinless density matrix $\mathbf{D}$ and the spin density matrix $\mathbf{P}^s$ defined as the sum and difference of matrices $\mathbf{P}^\alpha$ and $\mathbf{P}^\beta$

$$\mathbf{D} = \mathbf{P}^\alpha + \mathbf{P}^\beta , \qquad \mathbf{P}^s = \mathbf{P}^\alpha - \mathbf{P}^\beta , \tag{6.16}$$

the definition of the bond order can identically be written also as

$$B_{AB} = \sum_{\mu\in A}\sum_{\nu\in B}\left[(\mathbf{D}\mathbf{S})_{\mu\nu}(\mathbf{D}\mathbf{S})_{\nu\mu} + (\mathbf{P}^s\mathbf{S})_{\mu\nu}(\mathbf{P}^s\mathbf{S})_{\nu\mu}\right] . \tag{6.17}$$

In the most common case of a singlet system treated at the RHF level the spin density vanishes and this expression reduces simply to

$$B_{AB}^{closed\ shell} = \sum_{\mu\in A}\sum_{\nu\in B}(\mathbf{D}\mathbf{S})_{\mu\nu}(\mathbf{D}\mathbf{S})_{\nu\mu} . \tag{6.18}$$

This formula was first introduced by Giambiagi et al. [65] for the semiempirical "extended Hückel theory," by considering the idempotency properties of the density operator. They also stressed that *the bond order index reduces to the Wiberg index in the case of an orthonormalized basis* ($\mathbf{S} = \mathbf{1}$). Due perhaps to the use of unusual tensorial notations, their work, however, did not receive the attention it deserved, and had, therefore, a limited impact on the development of the theory.

Until recently [66] we had considered that the definitions (6.15)–(6.18) are applicable in the general, correlated case, too. (This point of view was expressed in, e.g., [64, 67].) As will be discussed later, that is not quite the case, and an improvement needs to be introduced—although it results in only marginal differences of the numerical results. In accord with that, except when otherwise stated, we shall concentrate on the single determinant wave functions: anyway, in most cases they carry the chemically relevant information—leaving out, of course, the cases in which very accurate numbers are required.

Invariance

The bond order index can be considered the simplest diatomic quantity that is quadratic in the interatomic elements of the matrices $\mathbf{P}^{\alpha}\mathbf{S}$ and $\mathbf{P}^{\beta}\mathbf{S}$ (or $\mathbf{DS}$, $\mathbf{P}^{s}\mathbf{S}$) and has the correct rotational–hybridizational invariance. In that sense it can be considered as a next logical step after Mulliken's overlap population, which is the simplest invariant diatomic quantity linear in the D-matrix elements. A detailed proof of the invariance of the bond order has been given in [5], so here we give only a brief outline of it.

As already discussed (c.f. pages 40, 45), we may treat the rotational–hybridizational transformations of the basis orbitals of all atoms simultaneously by introducing the transformation of the basis with a block-diagonal matrix $\mathbf{T}$ containing the blocks $\mathbf{T}^{A}$ corresponding to the individual atoms along its diagonal. Applying this transformation and using equalities (4.45) and (4.42) we have:

$$
\begin{aligned}
\sum_{\mu\in A}\sum_{\nu\in B}(\mathbf{P}'\mathbf{S}')_{\mu\nu}(\mathbf{P}'\mathbf{S}')_{\nu\mu} &= \\
&= \sum_{\mu\in A}\sum_{\nu\in B}\left[\mathbf{T}^{-1}\mathbf{P}(\mathbf{T}^{\dagger})^{-1}\mathbf{T}^{\dagger}\mathbf{S}\mathbf{T}\right]_{\mu\nu}\left[\mathbf{T}^{-1}\mathbf{P}(\mathbf{T}^{\dagger})^{-1}\mathbf{T}^{\dagger}\mathbf{S}\mathbf{T}\right]_{\nu\mu} \\
&= \sum_{\mu\in A}\sum_{\nu\in B}(\mathbf{T}^{-1}\mathbf{PST})_{\mu\nu}(\mathbf{T}^{-1}\mathbf{PST})_{\nu\mu} \qquad (6.19)\\
&= \sum_{\mu,\rho,\lambda\in A}\;\sum_{\nu,\tau,\eta\in B}(\mathbf{T}^{-1})_{\mu\rho}(\mathbf{PS})_{\rho\tau}T_{\tau\nu}(\mathbf{T}^{-1})_{\nu\eta}(\mathbf{PS})_{\eta\lambda}T_{\lambda\mu} \\
&= \sum_{\rho\in A}\sum_{\tau\in B}(\mathbf{PS})_{\rho\tau}(\mathbf{PS})_{\tau\rho}\,,
\end{aligned}
$$

proving the invariance. Here $\mathbf{P}$ is either of the matrices $\mathbf{P}^{\alpha}$, $\mathbf{P}^{\beta}$, $\mathbf{D}$, or $\mathbf{P}^{s}$, and we have taken into account that because $\mu\in A$, $(\mathbf{T}^{-1})_{\mu\rho}$ and $T_{\lambda\mu}$ differ from zero only if one has also $\rho\in A$, and $\lambda\in A$, respectively. Similar considerations restrict subscripts τ, η to atom B. Then the summations over μ and ν give the Kronecker deltas $\delta_{\lambda\rho}$ and $\delta_{\tau\eta}$.

Asymptotic exchange energy component and the bond order index

Proceeding in a close analogy with the Wiberg index of the semiempirical CNDO scheme, one would like also to connect the *ab initio* bond order index with a respective diatomic exchange energy component. The latter can be, in principle, calculated on the basis of the diatomic part of the exchange density $\varrho_2^x(\vec{r},\vec{r}\,')$, by introducing the term $1/|\vec{r}-\vec{r}\,'|$ describing the Coulomb interaction between the electrons and integrating the relevant diatomic term on the right-hand side of Eq. (3.48), as

$$\begin{aligned} E_{AB}^x &= -\sum_{\mu\in A}\sum_{\nu\in B}\sum_{\tau,\rho}\left(P_{\mu\tau}^{\alpha}P_{\nu\rho}^{\alpha}+P_{\mu\tau}^{\beta}P_{\nu\rho}^{\beta}\right)\iint\frac{\chi_\tau^*(\vec{r})\chi_\mu(\vec{r}\,')\chi_\rho^*(\vec{r}\,')\chi_\nu(\vec{r})}{|\vec{r}-\vec{r}\,'|}dvdv' \\ &= -\sum_{\mu\in A}\sum_{\nu\in B}\sum_{\tau,\rho}\left(P_{\mu\tau}^{\alpha}P_{\nu\rho}^{\alpha}+P_{\mu\tau}^{\beta}P_{\nu\rho}^{\beta}\right)[\tau\rho|\nu\mu]\,. \end{aligned} \tag{6.20}$$

Here the minus sign appears because, as shown in Eq. (3.41), the exchange density $\varrho_2^x(\vec{r},\vec{r}\,')$ enters the total pair density $\varrho_2(\vec{r},\vec{r}\,')$—and thus the energy—with a minus sign.* This is an interesting expression; but, unlike the CNDO case, there is no strict proportionality between this quantity and the bond order index B_{AB}.

The two-electron integral $[\tau\rho|\nu\mu]$ describes the electrostatic interaction of the charge densities $\chi_\tau^*(\vec{r})\chi_\nu(\vec{r})$ and $\chi_\rho^*(\vec{r}\,')\chi_\mu(\vec{r}\,')$ which are attributed for our purpose to atoms B and A, respectively, because orbitals χ_ν and χ_μ coming from the "ket" are centered on them. The overall value of the interacting charges is described by the overlap integrals $S_{\tau\nu}=\int\chi_\tau^*(\vec{r})\chi_\nu(\vec{r})dv$ and $S_{\rho\mu}=\int\chi_\rho^*(\vec{r}\,')\chi_\mu(\vec{r}\,')dv'$. If atoms A and B are far enough apart, this interaction can be subjected to a multipolar expansion, the first term of which describes the interaction of the point-like charges of values $S_{\tau\nu}$ and $S_{\rho\mu}$ at the distance which may be approximated by the interatomic distance R_{AB}:

$$[\tau\rho|\nu\mu]\sim\frac{S_{\tau\nu}S_{\rho\mu}}{R_{AB}}\,. \tag{6.21}$$

Thus we get the asymptotic expression

$$E_{AB}^x\sim-\sum_{\mu\in A}\sum_{\nu\in B}\sum_{\tau,\rho}\left(P_{\mu\tau}^{\alpha}P_{\nu\rho}^{\alpha}+P_{\mu\tau}^{\beta}P_{\nu\rho}^{\beta}\right)\frac{S_{\tau\nu}S_{\rho\mu}}{R_{AB}}\,, \tag{6.22}$$

*When the inter-electronic interaction of a charge density with itself is calculated, a factor of 1/2 should be introduced to avoid double counting; it is compensated by omitting the term of Eq. (3.41) in which A and B are interchanged: they would give a factor of 2 as they differ only by interchanging the integration variables $\vec{r}$ and $\vec{r}\,'$.

i.e., performing the summation over subscripts τ, ρ and using the definition (6.15) of the bond order index:

$$E^x_{AB} \sim -\frac{1}{2R_{AB}} B_{AB} . \tag{6.23}$$

This asymptotic expression is the direct *ab initio* analogue of Eq. (6.10) obtained for the CNDO theory. (One has also $\gamma_{AB} \sim 1/R_{AB}$ at large distances.) Originally, the expression of the asymptotic exchange energy component was first obtained (in a more sophisticated manner using the "chemical Hamiltonian approach" [26]) and then that was used as a starting point to get the definition (6.15) of the bond order index [24].*

6.4 Ab initio bond order indices of homonuclear diatomics

In the ab initio case the behavior of the bond order index (6.15) is similar to that of the Wiberg index in the CNDO frame: it gives values which are close to the classical bond multiplicities and often correlate well with the changes in the strengths of similar bonds.†

In the special case of homonuclear diatomics they give values coinciding—or very close—to the classical integer values, if a minimal basis set is used, while somewhat larger deviations occur for larger basis sets. Owing to the presence of the overlap, the derivation of Borisova and Semenov discussed above cannot be directly applied, so we had to look to a generalization of it [67,68].

We will again restrict our consideration to a given pair of hybrids χ^A_μ and χ^B_μ on the two atoms, from which one can form a bonding and an antibonding MO; they are with a proper normalization:

$$\begin{aligned} \varphi^b_\mu &= \frac{1}{\sqrt{2(1+s_\mu)}}(\chi^A_\mu + \chi^B_\mu) , \\ \varphi^a_\mu &= \frac{1}{\sqrt{2(1-s_\mu)}}(\chi^A_\mu - \chi^B_\mu) . \end{aligned} \tag{6.24}$$

It is assumed that the phases of the basis orbitals are selected to provide the overlap integral $s_\mu = \langle\chi^A_\mu|\chi^B_\mu\rangle$ to be positive; if it happens to be negative,

*Paper [24] contained a misprinted coefficient (2 instead of 1/2). Also, the final definition of the bond order index for *open shell systems* was given only somewhat later [25,63,64].

†Part of this section is reprinted from my paper [67] with permission from Wiley.

the notations of the bonding and antibonding orbitals φ_μ^b and φ_μ^a should be interchanged.

Now we consider single determinant wave functions built up of these orbitals filled with the respective spins $\sigma = \alpha$ or β. We assume that the order of the basis orbitals is selected as χ_{a1}, χ_{b1}, χ_{a2}, χ_{b2} and so on; then the density matrices $\mathbf{P}^\sigma$ are block-diagonal with 2 by 2 non-zero blocks on the main diagonal. If the overlap matrix is also assumed to have a block-diagonal structure, i.e., the basis orbitals on the individual atoms are orthonormalized and their interatomic overlaps have the "pairing" property

$$\langle \chi_\mu^A | \chi_\nu^B \rangle = s_\mu \delta_{\mu\nu} \tag{6.25}$$

then it is again sufficient to consider only 2 by 2 blocks of the respective matrices at a time. In such a case the bond order index (6.15) will represent a sum of the contributions originating from the individual pairs of basis orbitals χ_μ^A, χ_μ^B, and we may study them independently of each other. We shall denote the respective blocks of the matrices $\mathbf{S}$ and $\mathbf{P}^\sigma$ as $\mathbf{S}_\mu$ and $\mathbf{P}_\mu^\sigma$, respectively.

Let us first consider the case when the bonding orbital φ_μ^b is occupied with spin σ in the wave function, but its antibonding counterpart is not. In that case one readily gets

$$\mathbf{P}_\mu^\sigma = \frac{1}{2(1+s_\mu)} \begin{pmatrix} 1 & 1 \\ 1 & 1 \end{pmatrix} \tag{6.26}$$

and the matrix product $\mathbf{P}_\mu^\sigma \mathbf{S}_\mu$ is

$$\mathbf{P}_\mu^\sigma \mathbf{S}_\mu = \frac{1}{2(1+s_\mu)} \begin{pmatrix} 1 & 1 \\ 1 & 1 \end{pmatrix} \begin{pmatrix} 1 & s_\mu \\ s_\mu & 1 \end{pmatrix} = \tfrac{1}{2} \begin{pmatrix} 1 & 1 \\ 1 & 1 \end{pmatrix} \tag{6.27}$$

It follows from this result that $(\mathbf{P}_\mu^\sigma \mathbf{S}_\mu)_{12} (\mathbf{P}_\mu^\sigma \mathbf{S}_\mu)_{21} = \frac{1}{4}$, and by substituting into the definition (6.15) we see that the orbital φ_μ^b contributes $\frac{1}{2}$ to the bond order B_{AB} if it is occupied once. (There is a common factor of 2 in the equation.) If this bonding orbital is doubly occupied, its contribution to the bond order index is unity: we get a contribution of $\frac{1}{2}$ for both spins $\sigma = \alpha$ *and* β.

One can see that the same result is also obtained if the antibonding orbital φ_μ^a is singly or doubly occupied (but φ_μ^b is empty). Although for the ground state of a homonuclear diatomics that can hardly be the case, this fact indicates that the bond order indices (unlike the energy components to be discussed in Chapter 8) are not always able to distinguish between the bonding and antibonding situations.

If both orbitals φ_μ^b and φ_μ^a are occupied with the spin σ, then we get for matrix $\mathbf{P}_\mu^\sigma$

$$\mathbf{P}_\mu^\sigma = \frac{1}{1-s_\mu^2}\begin{pmatrix} 1 & -s_\mu \\ -s_\mu & 1 \end{pmatrix} \tag{6.28}$$

and for $\mathbf{P}_\mu^\sigma \mathbf{S}_\mu$

$$\mathbf{P}_\mu^\sigma \mathbf{S}_\mu = \begin{pmatrix} 1 & 0 \\ 0 & 1 \end{pmatrix} \tag{6.29}$$

Therefore, if both the bonding and antibonding combinations of a pair of corresponding orbitals are occupied with a given spin, then there is a complete cancellation and the given pair of corresponding orbitals does not contribute to the bond order index B_{AB}.

These results indicate that in the case of homonuclear diatomics the *sufficient condition* of getting a half-integer or integer value for the bond order index is that the occupied orbitals are built up of hybrid orbitals satisfying the "pairing" condition (6.25). Although in many cases that is ensured automatically by symmetry, it is to be noted that the conditions of the derivation were exactly fulfilled—even if a minimal basis is used—only if one could neglect completely the interaction (overlap) of the core orbitals, actually the $1s$ orbitals (K shell) for the first row atoms, with all the orbitals of the partner atom. In that case the $1s$ atomic orbitals would not be mixed with any other orbitals in the pairing procedure, and their sum and difference would coincide with the $1\sigma_g$ and $1\sigma_u$ canonical molecular orbitals. As the atomic number increases in the series from Li_2 to F_2, the $1s$ orbitals become relatively more compact and more separated from the valence shells, and thus the bond orders deviate less from the ideal integer values (c.f. Table 6.1).

If one looks at the canonical orbitals of the diatomics, calculated at the minimal basis level, one does not always see the pairs of orbitals like in Eq. (6.24), as far as the σ-orbitals are concerned, but that is not a real problem, except in the case of C_2. The point is that there are two valence AOs of σ symmetry (one s, one p_σ on each atom) out of which one can construct two bonding and two antibonding MOs. For N_2, O_2, and F_2 both bonding combinations are occupied, therefore *any* bonding σ-orbital is in the occupied space, including the bonding counterpart of the actually occupied antibonding canonical orbital. One can, therefore, perform a unitary transformation between the occupied orbitals, and express the wave function in terms of a pair of σ-orbitals of type Eq. (6.24), giving no contribution to the bond order, and a bonding σ-orbital the antibonding counterpart of which is empty, giving a contribution of one to the bond order. In C_2, however, one bonding and one antibonding σ-type valence MOs are occupied, which are not the counterpart of each other, so their contributions do not compensate, which leads to a bond order value which is far from an integer value.

Table 6.1
Bond orders of singlet homonuclear diatomics calculated by STO-6G, and cc-pVTZ basis sets at the equilibrium bond distances

Molecule	Bond order	
	STO-6G	cc-pVTZ
H_2	1.0000	1.0000
Li_2	0.9980	1.0065
Be_2^a	1.9987	2.0111
B_2	2.9994	3.0087
C_2	3.3328	3.3571
N_2	3.0000	2.9198
O_2^b	2.0000	1.8116
F_2	1.0000	0.9109

*STO-6G data: [67] permission from Wiley.
[a] For Be_2 there are two close lying states, one of Σ, another of Π type, at both HF and CAS-SCF levels of theory. The data in the table refer to the Π state having a minimum on the potential curve. (The Σ curve is repulsive everywhere and exhibits continuously changing bond orders.)
[b] For the triplet ground state of O_2 the bond orders are 2.0000 and 1.8296 for the two basis sets considered.

One can see that the bond orders calculated by using the sufficiently large cc-pVTZ basis set—containing 14 basis functions on the hydrogen and 30 basis functions on each second row atom (c.f. page 57)—do not differ too much from the ideal integer values. The number of effective AOs having occupation numbers significantly differing from zero is only five in these atoms, similar to the other examples discussed in Section 5.2, so the deviation from the ideal integer values should be mainly attributed to the small deviations of the structure of the bonding and antibonding orbitals, destroying the subtle balance leading to the exact compensation in Eq. (6.29).

The bond orders for polyatomic molecules are not strictly integers even in the simplest cases. Nonetheless, as the data for a set of prototype molecules shown in Table 6.2, they are usually close to the values corresponding to the "chemical expectations." It is evident from the data in Tables 6.1 and 6.2 that the bond order values depend on the basis set used and, therefore, it is meaningful only to compare bond order values calculated by using strictly the same basis set.

6.5 Bond orders in three-center bonds

When considering the electronic structure of stable molecules, chemists usually think in terms of electron pairs occupying core or lone-pair orbitals or orbitals responsible for the chemical bonds. The bonding electron pairs usually fall into two broad groups: they correspond either to localized bonds between a pair of adjacent atoms or belong to delocalized (e.g., π-electron) subsystems.* There are, however, systems in which the situation can be best described by postulating "two-electron three-center" (sometimes also "four-electron three-center") bonds; the best known example is the diborane molecule B_2H_6. In this molecule the two boron atoms are kept together with two "bridging" hydrogen atoms, each of which is connected with partial bonds to both borons *symmetrically*. (This symmetry distinguishes the structure of the diborane molecule from the usual hydrogen-bonded systems.) The bonds between the borons and the four "terminal" hydrogens can be considered two-electron two-center bonds, so they do not need any special discussion.

A peculiarity of these "two-electron three-center" bonds is that they not only describe bonding interactions between the adjacent atoms but, as a result of a sort of interference, they generate some bond order (i.e., attractive exchange interaction) between the boron atoms which are too far apart to have any significant direct interaction; this effect can significantly increase the stability of these systems. It may be worthwhile to study this problem by considering a small analytical model—the results of Chapter 5 about effective atomic orbitals indicate that such a simple model may be adequate even if the actual calculations use a larger basis.

Let χ_1, χ_2 and χ_3 be normalized atomic orbitals: χ_2 is the central hy-

*Admitting "resonance" between different "Kekulé-structures," usually the latter can also be described by means of two-center bonds.

Table 6.2
Bond orders of some prototype molecules calculated by using STO-3G, 6-31G**, and cc-pVTZ basis sets at the equilibrium bond distances

Molecule	Bond	Bond order		
		STO-3G	6-31G**	cc-pVTZ
Methane	C–H	0.991	0.978	0.977
Ethane	C–C	1.010	0.967	0.951
	C–H	0.984	0.977	0.982
Ethylene	C–C	2.020	1.969	1.887
	C–H	0.976	0.975	0.967
Acetylene	C–C	2.999	3.190	2.775
	C–H	0.985	0.893	0.932
Propane	C–C	1.001	0.970	0.952
	C_1–H (2×)	0.984	0.977	0.982
	C_1–H (1×)	0.984	0.978	0.987
	C_2–H	0.977	0.976	0.982
trans-Butadiene	C_1–C_2	1.944	1.896	1.818
	C_2–C_3	1.041	1.066	1.012
	C_1–H	0.975	0.975	0.969
		0.977	0.974	0.979
	C_2–H	0.968	0.967	0.956
Benzene	C–C	1.435	1.450	1.372
	C–H	0.972	0.963	0.979
Methyl radical	C–H	0.976	0.964	0.971
Water	O–H	0.964	0.884	0.998
Ammonia	N–H	0.962	0.918	0.984
CO	C–O	2.515	2.307	2.470
HCN	C–N	2.989	2.932	3.081
	C–H	0.968	0.892	0.928
CN^- ion	C–N	2.884	2.857	2.805
CN radical	C–N	2.162	2.506	2.781
Ethanol	C–C	0.997	0.980	0.964
	C–O	0.987	0.869	1.000
	C_1–H	0.966	0.969	0.976
	C_2–H (2×)	0.983	0.972	0.979
	(1×)	0.983	0.973	0.978
	O–H	0.950	0.878	1.004

drogenic orbital while χ_1 and χ_3 are the hybrid orbitals of the two borons oriented to the bridging hydrogen. It is assumed that the overlap of the boron and hydrogenic orbitals $\langle\chi_1|\chi_2\rangle = \langle\chi_2|\chi_3\rangle = S$, whereas the overlap of the two boron orbitals can be neglected. Then the overlap matrix is

$$\mathbf{S} = \begin{pmatrix} 1 & S & 0 \\ S & 1 & S \\ 0 & S & 1 \end{pmatrix} . \tag{6.30}$$

We construct the simplest unpolarized three-center orbital as

$$\psi = \frac{1}{\sqrt{2(1+\sqrt{2}S)}} \left[\chi_2 + \frac{1}{\sqrt{2}}(\chi_1 + \chi_3) \right] , \tag{6.31}$$

Using Eqs. (6.30) and (6.31), simple calculation gives for the matrices **D** and **DS** the result

$$\mathbf{D} = \frac{1}{2(1+\sqrt{2}S)} \begin{pmatrix} 1 & \sqrt{2} & 1 \\ \sqrt{2} & 2 & \sqrt{2} \\ 1 & \sqrt{2} & 1 \end{pmatrix} ; \tag{6.32}$$

$$\mathbf{DS} = \frac{1}{2} \begin{pmatrix} 1 & \sqrt{2} & 1 \\ \sqrt{2} & 2 & \sqrt{2} \\ 1 & \sqrt{2} & 1 \end{pmatrix} , \tag{6.33}$$

respectively. The value $(\mathbf{DS})_{22} = 1$ obtained shows that this bond is indeed unpolarized in the sense that the gross orbital population on the hydrogenic orbital equals one.

Using the matrix **DS** in (6.33), one easily obtains the bond order values $B_{12} = B_{23} = 1/2$; $B_{13} = 1/4$. (It is interesting that these values do not depend on the actual value of the overlap integral S.) As there are two such "two-electron three-center" bonds in the diborane molecule, a resulting boron-boron bond order of 1/2 can also be predicted. Table 6.3 shows that the predictions of this simple analytical model are applicable for a wide range of different basis sets.

"Four-electron three-center" bonds can be observed, for instance, in the axial bonds of the hypervalent sulfur molecules like SF_4. For such systems quite similar calculations can be performed as those described above; one has only to take into account that the central orbital is in this case an antisymmetric p-orbital; thus one of the overlap integrals equals S, another $-S$. Also, in these systems the non-bonding orbital proportional to the sum $\chi_1 + \chi_3$ is also doubly occupied. One gets exactly the same bond order values as for the "two-electron three-center" bond, which is not surprising in light of the fact that the second pair of electrons is placed on a non-bonding orbital. The

Table 6.3
Some bond orders of diborane molecule calculated by using STO-3G, 6-31G**, and cc-pVTZ basis sets at the equilibrium bond distances

Bond	Bond order		
	STO-3G	6-31G**	cc-pVTZ
B–H$_{br.}$	0.483	0.476	0.476
B–H$_{term.}$	0.988	0.993	0.998
B–B	0.506	0.470	0.525

reduced values of the bond orders explain completely why the axial bonds are longer than the conventional two-electron two-center equatorial ones. (Of course, for systems like SF_4 the polarization of the bonds can also have a significant importance.)

There is a quite different approach, too, to the problem of three-center—and, in general, multi-center—bonding. (In the following considerations we shall restrict ourselves to closed-shell systems.)

The normalization of the first-order density matrix is

$$\sum_A \sum_{\mu \in A} (\mathbf{DS})_{\mu\mu} = N \, . \tag{6.34}$$

where N is the number of electrons in the system. The sum over the orbitals belonging to a given atom A gives Mulliken's gross atomic population. The contribution of each basis orbital to this population—so to the sum (3.44)—is considerable if the orbital is significantly populated and is negligible if the given orbital is (nearly) empty.

Now, for a closed-shell system one has the idempotency relation $\mathbf{DS}^2 = 2\mathbf{DS}$. This means that one can write

$$\sum_{A,B} \sum_{\mu \in A} \sum_{\nu \in B} (\mathbf{DS})_{\mu\nu} (\mathbf{DS})_{\nu\mu} = 2N \, . \tag{6.35}$$

In this presentation, only those pairs of atoms $A \neq B$ give considerable contributions for which the bond order (6.18) is significant. Analogously, we may apply the same idempotency relation once again:

$$\sum_{A,B,C} \sum_{\mu \in A} \sum_{\nu \in B} \sum_{\rho \in C} (\mathbf{DS})_{\mu\nu} (\mathbf{DS})_{\nu\rho} (\mathbf{DS})_{\rho\mu} = 4N \, . \tag{6.36}$$

It appears that only those of different triads of atoms give a considerable contribution to the sum (6.36) which do exhibit a significant three-center bonding effect. These considerations may be generalized to higher multi-center bondings, of course.

6.6 The charge-fluctuation definition of bond order

In Section 4.2 we introduced the operator $\hat{N}_A$ of atomic population defined in the "mixed" second quantization framework:

$$\hat{N}_A = \sum_{\mu \in A} \hat{\chi}_\mu^+ \hat{\varphi}_\mu^- , \tag{6.37}$$

the expectation value of which is the Mulliken population of atom A. The expectation value of the product $\hat{N}_A\hat{N}_B$ of two such operators ($A \neq B$) however does not equal the products of the respective Mulliken populations, but also contains a term of exchange origin [26], which *for single determinant wave functions* is proportional to the bond order between the atoms (Appendix A.1):

$$\begin{aligned} \langle \hat{N}_A\hat{N}_B \rangle &= Q_A Q_B - \sum_{\mu \in A} \sum_{\nu \in B} \left[(\mathbf{P}^\alpha \mathbf{S})_{\mu\nu} (\mathbf{P}^\alpha \mathbf{S})_{\nu\mu} + (\mathbf{P}^\beta \mathbf{S})_{\mu\nu} (\mathbf{P}^\beta \mathbf{S})_{\nu\mu} \right] \\ &= Q_A Q_B - \tfrac{1}{2} B_{AB} \end{aligned} \tag{6.38}$$

(This is the result which originally led to the introduction of the bond order index [24] and this is how the relationship (6.23) was first obtained.)

The relationship (6.38) indicates that the emergence of the bond order index B_{AB} in the case of single determinant wave functions may be attributed to the fact that the expectation value of the operator product $\hat{N}_A\hat{N}_B$ differs from the product of the respective expectation values: $\langle \hat{N}_A\hat{N}_B \rangle \neq \langle \hat{N}_A \rangle \langle \hat{N}_B \rangle = Q_A Q_B$. In fact, equality (6.38) can be regrouped as

$$B^f_{AB} = -2\left(\langle \hat{N}_A\hat{N}_B \rangle - Q_A Q_B \right) \tag{6.39}$$

Superscript "f" is introduced to stress that this quantity is defined through the operators $\hat{N}_A$ of atomic population and their expectation values—Mulliken's gross atomic populations—Q_A. ("f" stands for "fluctuations," see below.)

Giambiagi et al. [32] have rewritten the relationship (6.39) to the equivalent form

$$B^f_{AB} = -2\left\langle \left(\hat{N}_A - Q_A \right) \left(\hat{N}_B - Q_B \right) \right\rangle \tag{6.40}$$

which permits a "statistical" interpretation of the bond order index: it measures the degree in which the *fluctuations* of the electron populations on the two atoms—i.e., their deviations from the mean (expectation) values—are *correlated* with each other: if the atoms are connected with a covalent bond, then the decrease of the electron density on one of the atoms involves the increase of it on the partner atom, and *vice versa.**

Of course, if one considers a non-polarized molecular orbital involving only two atoms, then the electron(s) on it are either on this atom or on that atom, and one gets a full correlation between the fluctuations of the charges on the two atoms in question. That orbital then gives rise to a contribution of one to the value of Eq. (6.40) if it is doubly occupied (1/2 if it is singly occupied); if there are several independent (not involving the same basis functions) unpolarized MOs on that pair of atoms, one gets the respective multiple bond orders. The extent to which the situation around the given pair of atoms corresponds to the presence of a given number of localized two-center orbitals is then reflected by the actual value of the bond order index.

6.7 Bond order in the correlated case

It is important to note that Equations (6.39), and (6.40) give results coinciding with the original definition in Eq. (6.15) for the single determinant wave functions only, but the equality $B^f_{AB} = B_{AB}$ does not hold for the general (correlated) wave functions. It would be tempting to use the "pictorial" fluctuation definition as the general one, but it leads to serious difficulties if correlated wave functions are concerned.†

To see the problem, let us consider the example of the H_2 molecule treated at the minimal basis level. Assuming that the basis orbitals centered on the two atoms A and B are χ_a and χ_b, the most general wave function having the symmetry of the exact ground state can be written as‡

$$\Psi(1,2) = \mathcal{N}\left[\hat{\mathscr{A}}[\chi_a^\alpha(1)\chi_b^\beta(2)] + \hat{\mathscr{A}}[\chi_b^\alpha(1)\chi_a^\beta(2)]\right. \tag{6.41}$$

*The right-hand sides of Eqs. (6.39) and (6.40) are identically equal because Q_A and Q_B are numbers and not operators, so $\langle Q_A\hat{N}_B\rangle = \langle\hat{N}_A Q_B\rangle = Q_A Q_B$.

†Part of this section is reprinted from my paper [66] with permission from Elsevier.

‡We use the fixed order of the spin functions, not that of the spatial orbitals, hence the sign "+" instead of the usual "−".

$$+c\left(\mathscr{A}[\chi_a^\alpha(1)\chi_a^\beta(2)]+\mathscr{A}[\chi_b^\alpha(1)\chi_b^\beta(2)]\right)\Big],$$

where

$$\mathscr{N}=\frac{1}{\sqrt{2[(c+S)^2+(1+cS)^2]}}, \tag{6.42}$$

is a normalization coefficient depending on the overlap integral $S=\langle\chi_a|\chi_b\rangle$ and the parameter c. It is easy to see that for $c=1$ Eq. (6.41) recovers the usual LCAO-MO wave function in which the bonding orbital $\frac{1}{\sqrt{2(1+S)}}(\chi_a+\chi_b)$ is doubly occupied, while for $c=0$ the classical Heitler–London wave function [69]—containing the "covalent components" only—is obtained. The best wave function of this type corresponds to an optimized value of the parameter c—that is the so-called Weibaum's wave function [70] and is equivalent to a "full CI" performed in the minimal basis set.

One can write the equivalent of Eq. (6.41) in second quantized formalism as

$$|\Psi\rangle=\mathscr{N}\left[\hat{\chi}_a^{\alpha+}\hat{\chi}_b^{\beta+}+\hat{\chi}_b^{\alpha+}\hat{\chi}_a^{\beta+}+c\left(\hat{\chi}_a^{\alpha+}\hat{\chi}_a^{\beta+}+\hat{\chi}_b^{\alpha+}\hat{\chi}_b^{\beta+}\right)\right]|0\rangle\,. \tag{6.43}$$

Now, it is easy to see that only the covalent terms of Eq. (6.43) survive the application of the operator product $\hat{N}_A\hat{N}_B$ present on the right-hand side of Eq. (6.39), giving

$$\hat{N}_A\hat{N}_B|\Psi\rangle=\mathscr{N}\left[\hat{\chi}_a^{\alpha+}\hat{\chi}_b^{\beta+}+\hat{\chi}_b^{\alpha+}\hat{\chi}_a^{\beta+}\right]|0\rangle\,, \tag{6.44}$$

from which the expectation value of $\hat{N}_A\hat{N}_B$ is

$$\langle\hat{N}_A\hat{N}_B\rangle=\frac{1+S^2+2cS}{(c+S)^2+(1+cS)^2}\quad. \tag{6.45}$$

This leads to the expression for the right-hand side of Eq. (6.39)

$$B_{AB}^f=2c\frac{c+2S+cS^2}{(c+S)^2+(1+cS)^2}\quad. \tag{6.46}$$

It is easy to see that Eq. (6.46) gives the value of 1 if $c=1$, irrespective of the value of S, in full agreement with the facts that $B_{AB}^f=B_{AB}$ for single determinants and that for the H_2 molecule the RHF wave function predicts a bond order of unity at all interatomic distances. At the same time, this formula gives $B_{AB}^f=0$ in the case of the Heitler–London wave function ($c=0$), although it is well known that this wave function is able to account semiquantitatively for the bond formation. (The Heitler–London description of the H_2

molecule was the birth of quantum chemistry as science, as it represented the first breakthrough in understanding the quantum mechanical nature of the chemical bond.) Also, one obtains [64] the very low value $B^f_{AB} = 0.39$ for Weinbaums's optimized (i.e., full CI) wave function. Therefore, the quantity B^f_{AB} defined in Eqs. (6.39), (6.40) *cannot be reasonably used as a bond order (multiplicity) index*—which does not mean that it is of no significance for characterization of the bonding.

In this situation it was proposed [64] to use the same Eqs. (6.15), (6.17) for the definition of the bond order in the correlated case, too. That was equivalent to writing the pair density in the form

$$\varrho_2(\vec{r},\vec{r}') = \varrho(\vec{r})\varrho(\vec{r}') - \varrho_2^x(\vec{r},\vec{r}') + \varrho_2^c(\vec{r},\vec{r}') , \qquad (6.47)$$

where one, somewhat artificially, separated out the exchange density $\varrho_2^x(\vec{r},\vec{r}')$ in the form defined through the matrices $\mathbf{P}^\alpha$, $\mathbf{P}^\beta$ in Eq. (3.43), like in the single determinant case, and the "cumulant" part $\varrho_2^c(\vec{r},\vec{r}')$ describing the electron correlation.* That means that we used only this "reconstructed" exchange density for calculating bond orders, while the cumulant part of the pair density is not utilized in any manner. This had (and has, see below) the advantage that one could work explicitly with the first-order density matrices only, and the bond orders were determined exactly in the same manner as in the case of single determinant wave functions.

An important observation concerning the use of Equations (6.15), (6.17) beyond the single determinant level was made by Alcoba et al. [71] who called attention to the fact that the presence of the spin-density matrix in Eq. (6.17) can lead to inconsistent results when different spin-projections corresponding to the same S value of an open-shell system are compared. For instance, if one considers a triplet system at the ROHF level, then the $S_z = 1$ state is described with a single determinant, for which the applicability of Eq. (6.17) containing the spin density is obvious. At the same time for the $S_z = 0$ component of the triplet (which is *not a single determinant*), one has zero spin-density, so Eq. (6.17) reduces to Eq. (6.18) and one gets (slightly) different bond order values. To avoid this ambiguity, Alcoba et al. suggested

---

*That is a particular case of writing the overall second-order density matrix as a sum of terms expressible through the first-order density matrix and the respective cumulant:

$$\varrho_2(\mathbf{x}_1,\mathbf{x}_2;\mathbf{x}_1',\mathbf{x}_2') = \varrho_1(\mathbf{x}_1;\mathbf{x}_1')\varrho_1(\mathbf{x}_2;\mathbf{x}_2') - \varrho_1(\mathbf{x}_1;\mathbf{x}_2')\varrho_1(\mathbf{x}_2;\mathbf{x}_1') + \varrho_2^c(\mathbf{x}_1,\mathbf{x}_2;\mathbf{x}_1',\mathbf{x}_2')$$

where $\mathbf{x}_i$ denotes the combination of spatial and spin coordinates of electron i. It is known that for single determinant wave functions (and only for single determinant wave functions) the second-order density matrix can be expressed *via* the first-order one, and $\varrho_2^c \equiv 0$.

calculating the values corresponding to the highest spin projection and using them also for the lower one(s). As in practice one usually performs the calculations right for that highest S_z projection, this problem did not have a serious practical impact.

Most recently, however, we have realized that there is a closely related problem which needs to be addressed [66]. (Its solution permits also to treat the problem just mentioned without resorting to the wave function of another spin projection.) The problem was actually discovered for the ethylene molecule treated at the CAS-SCF level, but of course similar effects could be observed for other correlated systems, too.

If one considers the dissociation of the ethylene molecule into two methylenes, then at the large C–C distances the system consists of two triplet subsystems coupled into an overall singlet. If one applies the CAS(4,4) method, then in the dissociated limit that is equivalent to treating the two fragments at the ROHF level with two electrons outside the closed shells in each methylene. The $S_z = 1$ component of such an ROHF wave function represents an open-shell single determinant for which Eq. (6.17) was originally derived. Therefore, the bond orders of the methylene moieties should be calculated with Eq. (6.17), including the term with the spin-density connected with the two unpaired electrons. However, when doing the calculation for the overall singlet system, then Eq. (6.18) was applied, which meant omission of the terms containing the spin density. In this particular case the difference is marginal (the C–H bond order is 0.94 instead of 0.96 in the actual CAS(4,4)/6-31G** calculation) and all the qualitative behavior of the bond order (and valence, *vide infra*) indices along the ethylene dissociation curve was quite correct [67], but that situation is, of course, rather inadequate from the conceptual point of view. Therefore, we had to look for an improvement of the definition for the correlated case, permitting to achieve a correct dissociation of bond orders. Fortunately, that was possible without recurring to the full second-order density matrix, which has a significant practical advantage.

To get the correction, we shall first consider some properties of the matrix $2\mathbf{DS} - (\mathbf{DS})^2$, that will be discussed more in detail in Section 7.1 It vanishes in the RHF case, while in the ROHF one it can be transformed as

$$\begin{aligned} 2\mathbf{DS} - (\mathbf{DS})^2 &= 2(\mathbf{P}^\alpha\mathbf{S} + \mathbf{P}^\beta\mathbf{S}) - (\mathbf{P}^\alpha\mathbf{S} + \mathbf{P}^\beta\mathbf{S})(\mathbf{P}^\alpha\mathbf{S} + \mathbf{P}^\beta\mathbf{S}) \\ &= 2(\mathbf{P}^\alpha\mathbf{S}\mathbf{P}^\alpha\mathbf{S} + \mathbf{P}^\beta\mathbf{S}\mathbf{P}^\beta\mathbf{S}) - (\mathbf{P}^\alpha\mathbf{S} + \mathbf{P}^\beta\mathbf{S})(\mathbf{P}^\alpha\mathbf{S} + \mathbf{P}^\beta\mathbf{S}) \\ &= \mathbf{P}^\alpha\mathbf{S}\mathbf{P}^\alpha\mathbf{S} + \mathbf{P}^\beta\mathbf{S}\mathbf{P}^\beta\mathbf{S} - (\mathbf{P}^\alpha\mathbf{S}\mathbf{P}^\beta\mathbf{S} + \mathbf{P}^\beta\mathbf{S}\mathbf{P}^\alpha\mathbf{S}) \\ &= \mathbf{P}^s\mathbf{S}\mathbf{P}^s\mathbf{S} \end{aligned} \tag{6.48}$$

where we have utilized the idempotency property of the single determinant density matrices $\mathbf{P}^\sigma\mathbf{S} = \mathbf{P}^\sigma\mathbf{S}\mathbf{P}^\sigma\mathbf{S}$ (σ=α *or* β.)

This result indicates that in the ROHF case matrix $\mathbf{P}^s\mathbf{S}$ can be considered in some sense a square root of matrix $2\mathbf{DS}-(\mathbf{DS})^2$. That is of particular importance, because the latter is a spin-free entity, which is *the same* for every S_z projection of the given open-shell state.

Unfortunately, the square root of a matrix is not unique, so one cannot calculate $\mathbf{P}^s$ directly from Eq. (6.48). However, in an orthonormalized (e.g., Löwdin) basis both matrices $2\mathbf{D}-\mathbf{D}^2$ and $\mathbf{P}^s$ are Hermitian and positive definite, and there is a Hermitian square root of a positive definite Hermitian matrix that can be considered privileged: that one which can be calculated by diagonalization and taking the *positive square root* of each eigenvalue—similarly to the procedure one usually applies for calculating $\mathbf{S}^{-1/2}$ when doing Löwdin-orthogonalization. This means that in the ROHF case one can reconstruct the spin-density matrix from the total density matrix by an appropriate procedure. Unfortunately, that is not the case for UHF wave functions.* Moreover, one can see that one can recover the spin-density matrix corresponding to the highest S_z projection of an open shell state, by using the matrix $2\mathbf{DS}-(\mathbf{DS})^2$, determined in a calculation performed for any S_z component.

In the correlated case one has $2\mathbf{DS}-(\mathbf{DS})^2 \neq 0$, too, even for a singlet. In that case the deviation is not connected with the spin-density but with correlation. But the $S_z = 0$ component of a triplet, etc., is not a single determinant either, so there is no big conceptual difference between such a state and a correlated wave function. Therefore, in a correlated singlet case one can introduce some "pseudo spin density matrix" $\mathbf{R}$, which is calculated from matrix $2\mathbf{DS}-(\mathbf{DS})^2$ exactly as the spin-density matrix is reconstructed in the ROHF case. That means writing

$$2\mathbf{DS}-(\mathbf{DS})^2 = \mathbf{RSRS} \ . \tag{6.49}$$

Then the improved formula of bond order becomes

$$B_{AB} = \sum_{\mu\in A}\sum_{\nu\in B}\left[(\mathbf{DS})_{\mu\nu}(\mathbf{DS})_{\nu\mu} + (\mathbf{RS})_{\mu\nu}(\mathbf{RS})_{\nu\mu}\right] \tag{6.50}$$

This formula can be considered quite general: in the open-shell (UHF or ROHF) single determinant case one should consider $\mathbf{R} = \mathbf{P}^s$. This formula permits to restore correct asymptotics of the bond order when a molecule

*In the ROHF case matrix $\mathbf{P}^s\mathbf{S}$ is also idempotent and, therefore, it is equal to $2\mathbf{DS}-(\mathbf{DS})^2$, too. However, we are looking for a scheme that can be generalized for the correlated systems.

dissociates into open-shell fragments and to get results independent of the S_z projection actually considered.

The actual calculation of matrix **R** can be accomplished as follows [66]. It can be easily checked (c.f. [5]) that when turning to the Löwdin-orthogonalized basis (superscript λ), matrix $\mathbf{2DS}-(\mathbf{DS})^2$ transforms to $\mathbf{Z}^\lambda = 2\mathbf{D}^\lambda - (\mathbf{D}^\lambda)^2$ as

$$\mathbf{Z}^\lambda = \mathbf{S}^{\frac{1}{2}}[\mathbf{2DS}-(\mathbf{DS})^2]\mathbf{S}^{-\frac{1}{2}} = \mathbf{S}^{\frac{1}{2}}[\mathbf{2D}-\mathbf{DSD}]\mathbf{S}^{\frac{1}{2}} \,. \tag{6.51}$$

One should diagonalize matrix $\mathbf{Z}^\lambda$, replace the eigenvalues in the diagonals by their square roots and back-transform to get matrix $(\mathbf{Z}^\lambda)^{1/2}$.* Then matrix **RS** can be obtained by turning back to the original basis as

$$\mathbf{RS} = \mathbf{S}^{-\frac{1}{2}}(\mathbf{Z}^\lambda)^{\frac{1}{2}}\mathbf{S}^{\frac{1}{2}} \,. \tag{6.52}$$

Let us return to the discussion of the H_2 minimal basis model considered above. The matrix elements $(\mathbf{P}^\sigma\mathbf{S})_{\mu\nu}$ ($\sigma = \alpha$ *or* β) necessary to determine the matrix elements $(\mathbf{DS})_{\mu\nu}$ occurring in the definition (6.50) of the bond order can be calculated [26] as the expectation values of the operator strings $\langle \hat{\chi}_\nu^{\sigma+} \hat{\varphi}_\mu^{\sigma-} \rangle$. It appears that the off-diagonal elements of matrix **RS** vanish for this model as a consequence of symmetry.† Thus in this particular case there is no correction due to that term, and one gets for the bond order

$$B_{AB} = 4\left[\frac{(c+S)(1+cS)}{(c+S)^2+(1+cS)^2}\right]^2 \,. \tag{6.53}$$

One can see that this definition gives $B_{AB} = 1$, as it should, irrespective of the value of S, for the case of $c = 1$, when the wave function reduces to the single LCAO-MO determinant, and that it predicts the bond order $[2S/(1+S^2)]^2 \sim 0.85$ for the Heitler–London wave function ($c = 0$). For the Weinbaum wave function with optimized exponent one gets [64] the quite

---

*This procedure is quite analogous to the calculation of $\mathbf{S}^{-1/2}$ when doing Löwdin orthogonalization. The eigenvalues of matrices $\mathbf{2DS}-(\mathbf{DS})^2$ and **Z** are the same and are equal to $n_i(2-n_i) \geq 0$ [72], where n_i are the occupation numbers of the natural orbitals ($0 \leq n_i \leq 2$). Therefore the calculation of the square roots cannot meet any problems.

†This is only true for minimal basis set calculations. For a full-CI calculation of H_2, performed in the cc-pVTZ basis (28 basis orbitals) at the interatomic distance of 0.74Å, there is a small correction of 0.014 due to the terms containing matrix **RS**, increasing the bond order from 0.952 to 0.966. We note in this connection that the corrections are usually that marginal. They are of importance not because of the numerical values but because of the correct dissociation properties.

reasonable bond order value of 0.947 for the equilibrium internuclear separation of the H_2 molecule. An interesting feature of this formula is that the coefficient c of the ionic terms and the overlap integral S enters it quite symmetrically, shedding some light on the physical significance of overlap in the bond formation.*

6.8 An example: Dimethylformamide

The bond order index Eq. (6.15), (6.17) is widely used in the literature to discuss quite different systems and problems. Here we shall consider a single example—the dimethylformamide molecule. (In the next section a special application will also be discussed.)

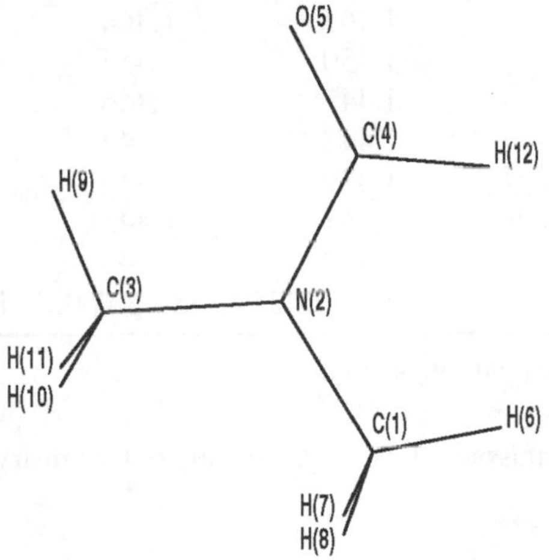

Figure 6.1
Numeration of atoms in the N,N-dimethylformamide molecule. [73] with permission of the Royal Society of Chemistry.

N,N-dimethylformamide $(CH_3)_2N\text{-}CHO$, Fig. 6.1, is a very simple molecule exhibiting, nevertheless, a rich chemistry. According to gas-phase elec-

*We recall in this connection that for two electrons on two *singly occupied orthogonal orbitals* the triplet has lower energy than the singlet (Hund's rule) and in an orthogonal basis only the admixture of the ionic terms lowers the energy of the singlet beyond that of the triplet. That also means that by using an orthogonal basis no chemical bonds can be described without the inclusion of the ionic terms in the wave function.

Table 6.4
Selected geometrical data of the N,N-dimethylformamide molecule (Ångströms and degrees)

Basis	$R(H_9 \cdots O)$	$\angle C_3NC_4$	$\angle C_1NC_4$
STO-3G[a]	2.359	119.9	121.8
6-31G	2.365	120.7	121.9
6-31G**	2.353	120.5	121.9
6-311G**	2.360	120.7	121.8
6-311G(2d,2p)	2.355	120.7	121.7
6-311++G(2d,2p)	2.366	120.9	121.7
cc-pVTZ	2.366	120.9	121.7
exp.[b]	2.40±0.03	120.8±0.3	122.3±0.4

Basis	$R(NC_1)$	$R(NC_3)$	$R(NC_4)$
STO-3G[a]	1.463	1.464	1.410
6-31G	1.450	1.454	1.349
6-31G**	1.442	1.446	1.349
6-311G**	1.443	1.447	1.348
6-311G(2d,2p)	1.439	1.444	1.344
6-311++G(2d,2p)	1.441	1.445	1.343
cc-pVTZ	1.439	1.444	1.344
exp.[b]	1.449±0.002	1.454±0.002	1.391±0.007

[a]Planar (saddle point) geometry.
[b]Ref. 74
*[73] with permission of the Royal Society of Chemistry.

tron diffraction data [74], dimethylformamide is planar or at least nearly planar, indicating the presence of a "resonance" between the carbonyl group and the nitrogen lone pair — essentially the formation of a small π-electron system; obviously the methyl groups are also connected to it with hyperconjugation.

Experiment revealed that the C-N-C angle which is *syn* to the carbonyl group is somewhat smaller than the other one, indicating that there must be some attractive C–H···O interaction, despite the relatively large (2.40 ± 0.03Å) H···O distance. This means that dimethylformamide is one of the simplest molecules in which the existence of a C–H···O interaction has been postulated on the basis of direct experimental results [74].

Table 6.5
Selected bond orders of the N,N-dimethylformamide molecule

Basis	C_4-O	N-C_1	N-C_3	N-C_4	O$\cdots H_9$
STO-3G[a]	1.913	0.962	0.959	1.041	0.004
6-31G	1.723	0.793	0.769	0.977	0.018
6-31G**	1.798	0.894	0.868	1.091	0.020
6-311G**	1.889	0.931	0.900	1.166	0.033
6-311G(2d,2p)	1.898	1.011	0.968	1.213	0.039
6-311++G(2d,2p)	1.636	0.943	0.909	1.183	0.035
cc-pVTZ	2.020	0.983	0.937	1.216	0.034

[a]Planar (saddle point) geometry.
*[73] with permission of the Royal Society of Chemistry.

In accord with the experiment, the molecule has been found completely planar in all SCF calculations, except when the minimal (STO-3G, 6G) basis was used, and no pyramidality has been observed at the MP2 or DFT levels either. This planarity is clearly to be assigned to the conjugation with the carbonyl group: trimethylamine is obtained strongly pyramidal. Figure 6.1 shows the numbering of atoms applied. Table 6.4 summarizes some of the geometrical data which seem to be of interest, calculated for several basis sets; Table 6.5 shows selected bond orders. The results are in good agreement with the notion of a delocalized π-electron system mentioned above: e.g., the N–C(O) distance is significantly shorter than the N–C(Me) distances, and the respective bond order significantly exceeds one. The N–C(Me) bonds are essentially single ones.

The C-N-C angle in *syn* position is consistently smaller than that in the *anti* position, in perfect agreement with the experiment. Remarkable also is the agreement with the experiment concerning the small but consistent difference between the N-C(Me) bond lengths indicating that the bond in *syn* position is slightly stretched, probably to provide better conditions for the C–H$\cdots$O interaction. The bond order values also are in full agreement with this: the shorter bond has a larger bond order.

The closest H$\cdots$O distance has been found 2.35-2.36 Å, and is practically independent of the basis set applied. This distance is slightly, but not significantly, shorter than the experimental one. Thus we may claim that these results confirm the existence of some attractive C–H$\cdots$O interaction, although the angle C–H$\cdots$O is only about 102–103°. (These geometrical features are present also at the DFT and MP2 levels; the H$\cdots$O distance even reduces a little.) This conclusion is confirmed by the presence of the partial H$\cdots$O bond order at any level of calculations applied—except the minimal basis STO-3G which is apparently not capable to describe such a fine effect.

Of course, when doing Hilbert-space analyses, one has to keep in mind that the results are basis dependent, and do not have, in general, any definite basis set limit. Therefore, when discussing any actual chemical problem, only results obtained by using strictly the same basis set should be compared. Also, it is obvious that only basis sets of sufficiently "atomic character" must be considered. Thus diffuse basis functions lacking true atomic nature may spoil the analysis. (In our case there is only one such basis set—the 6-31++G(2d,2p) one—and it indeed shows a deviation from the smooth change of results observed for the other basis sets.) Remarkably good is the performance of the simplest STO-3G basis, giving very "chemical" numbers, even if it cannot account for secondary interactions like H···O. The split shell 6-31G basis is known for some "over-polarization" of the bonds, resulting in too low bond orders. (The same is characteristic for the related, once very popular 4-31G basis set, too.) The further improvement of the basis led to numbers corresponding more and more to the chemical picture of the molecule—except the case of using diffuse functions, as already mentioned. The results given by the pretty large, but very well balanced and very atomic cc-pVTZ basis seems to be simply perfect.

6.9 An application: Predicting primary mass spectrometric cleavages

As noted in the introduction, the bond order indices and other parameters obtained in the *a posteriori* analysis may sometimes be used not only for interpretations but also for predictive purposes. This is the case if the molecule undergoes some change and we compare the bond orders (or other characteristics) of a given bond obtained before and after that change, without recalculating its length—then we can identify "intrinsic" weakenings or strengthenings of the bond, without an interplay with the changes in the bond length. Vertical ionization of a molecule under the effect of an electron impact may be considered a very characteristic change of this type. The existence of similar patterns in the mass-spectra of the similar molecules indicates that the highly complicated dynamical processes taking place in a mass spectrometric apparatus are governed to a large extent by some static properties related to the electronic structure of the neutral parent molecules and their positive ions formed upon electron impact. It appears that for molecules undergoing simple bond cleavage, the weakening of a given bond upon ionization may usually be attributed a decisive role for the whole process. This we have first observed by using the simple MNDO level of theory [75], and that was confirmed at the *ab initio* level, as well [76]. One has simply to compare the bond orders of the ionized species with those of the neutral parent molecule: one may expect the cleavage of those bonds for which a significant decrease of the bond order takes place upon ionization. (Similarly, one can also compare the respective energy components discussed in Section 8.4 below.)

From the practical point of view, one encounters the following complication: in many cases the fragmentation occurs after an electron being removed not from the HOMO but from one of the lower-lying orbitals. This is usually the case for aromatic molecules, the HOMO and the next-to-HOMO orbitals which are located on the aromatic ring; deleting an electron from these orbitals either does not lead to a decomposition or initiates more complex processes than simple bond cleavages. (Usually the mass spectra of such molecules do contain a dominating peak corresponding to the non-decomposed molecular ion, while the latter is often absent or is rather weak for molecules lacking aromatic moieties.)

Trying to do calculations for such systems we have encountered the difficulty that usually one cannot obtain converged SCF solutions for such "hole states" (i.e., states in which an electron is removed from some lower lying orbital), but then it was realized that there is no actual need to use them. For the purpose of qualitative prediction of mass spectra it is sufficient to apply the "quasi-Koopmans" approach, in which one simply uses the orbitals obtained for the neutral parent molecule and changes only the occupation numbers of them, by filling one of the MOs of the molecule with only one, instead of two, electrons.

Calculations of such type can be carried out very easily by using our freely available programs [77, 78]. Their result have the advantage that the qualitative conclusions are rather insensitive to the level of theory or the basis set applied. Table 6.6 compares the changes of bond orders in the "quasi-Koopmans" ionization from the HOMO of some aliphatic alcohols by using a minimal and a much bigger basis set. These data explain that the electron impact mass spectra of all these molecules is dominated by a peak with a mass-to-charge ratio z/m equal to 45, corresponding to the CH_3CHOH moiety.*

*There is some weakening of the C2–H, bonds, too, which does not manifest significantly in the mass spectra. This is in line with the general tendency that simple C–H bond cleavages are observed only rarely; for a further discussion we refer to [76].

Similarly, one may consider the effect of other changes in the molecule, too: for instance, comparing the bond orders of a protonated peptide chain with those of the non-protonated one, one can predict the decomposition patterns taking place in the tandem mass spectrometry (collision induced decomposition) [79].

Table 6.6
Bond order changes in the "quasi-Koopmans" ionization from the HOMO of some aliphatic alcohols calculated by using minimal (STO-3G) and cc-pVTZ (pVTZ) basis sets

	2-propanol		*2-butanol*		*2-pentanol*	
Bond	STO-3G	pVTZ	STO-3G	pVTZ	STO-3G	pVTZ
C1-C2						
neutral	0.988	0.937	0.993	0.988	0.988	0.975
HOMO	0.985	0.937	0.994	0.988	0.985	0.979
C2-C3						
neutral	0.968	0.901	0.969	0.963	0.968	0.958
HOMO	*0.827*	*0.701*	*0.765*	*0.758*	*0.827*	*0.755*
C2-O						
neutral	0.977	0.761	0.860	0.849	0.977	0.915
HOMO	1.158	0.961	1.054	1.040	1.158	1.158
C2-H						
neutral	0.959	0.952	0.955	0.971	0.959	0.995
HOMO	0.906	0.878	0.869	0.887	0.906	0.903
C3-C4						
neutral	—	—	0.976	0.967	0.992	0.951
HOMO	—	—	0.981	0.972	0.991	0.962
C4-C5						
neutral	—	—	—	—	1.000	0.963
HOMO	—	—	—	—	0.984	0.922

The table compares the bond orders calculated for the neutral molecule with those obtained by deleting one electron from the HOMO *without recalculating the orbitals for the ion* (quasi-Koopmans scheme). The numbers in *italics* indicate the bond the cleavage of which can be expected.

*[76] with permission from Elsevier.

6.10 Valences and free valences

The concept of valence indices also goes back in some sense to the same paper of Wiberg in which the Wiberg index had been introduced [58]. Wiberg made the remark that the quantity

$$b_\mu = 2q_\mu - q_\mu^2 , \tag{6.54}$$

where q_μ is the electron population on orbital χ_μ measures the extent to which that orbital can actually participate in chemical bond formation, because one has $b_\mu = 0$ for an empty orbital ($q_\mu = 0$) as well as for a doubly filled one ($q_\mu = 2$) having no chemical significance (e.g., a core orbital) and b_μ has a maximum equal to one for a singly occupied orbital. The sum of quantities b_μ for a given atom then could be used to measure the overall bonding ability of that atom, that is its *valence*. However, that sum is not invariant under rotational, etc., transformations, so it is not appropriate. However, it is easy to find a proper generalization of this sum, which is invariant. In the case of an orthogonal basis (e.g., the CNDO model) one may assume that the basis in which the sum of b_μs is calculated is the natural hybrid basis in which the atomic block of the spinless density matrix **D** is diagonal, $D_{\mu\nu} = D_{\mu\mu}\delta_{\mu\nu}$. In this basis one has

$$\sum_{\mu\in A} q_\mu^2 = \sum_{\mu\in A} D_{\mu\mu}^2 = \sum_{\mu,\nu\in A} D_{\mu\nu}D_{\nu\mu} , \tag{6.55}$$

and the quantity on the right-hand side does have the required invariance. (This can be proved exactly as the invariance of the Wiberg index discussed above.) Now, we can do the summation of the quantities b_μ pertinent to the basis orbitals of the given atom A with this generalization of the second term in Eq. (6.54) and get the *valence* of the atom in the framework of the CNDO formalism* (or, in general, for any theory which applies an orthonormalized basis set) [59, 80].

$$V_A^{\text{CNDO}} = 2Q_A - \sum_{\mu,\nu\in A} D_{\mu\nu}D_{\nu\mu} , \tag{6.56}$$

*An interesting coincidence was that the two groups proposed this definition almost simultaneously: Borisova and Semenov have a priority of some two weeks. But unfortunately their papers [59.62] were published in a hard-to-find journal and did not receive the attention they deserved.

as the sum of orbital populations q_μ is just the total electron population Q_A of the atom.

In the general case of a non-orthogonal basis, the role of the orbital population q_μ is attributed to Mulliken's gross orbital population $(\mathbf{DS})_{\mu\mu}$ and the proper invariant generalization of the sum of q_μ^2-s can also be done through the elements of the matrix **DS**. Thus we obtain [24] the general definition of the actual total valence of an atom in a molecule as

$$V_A = 2Q_A - \sum_{\mu,\nu\in A} (\mathbf{DS})_{\mu\nu}(\mathbf{DS})_{\nu\mu} \,. \tag{6.57}$$

with Q_A defined as Mulliken's gross atomic population of atom A.

In the RHF case there is an important inter-relation between valences and bond orders: *if doubly occupied orbitals are used, the valence of an atom equals the sum of its bond orders formed with all the other atoms.* This result is a consequence of the idempotency relationship $\mathbf{DS}^2 = 2\mathbf{DS}$ valid in the closed shell case. In fact, one may write for the term $2Q_A$ on the right-hand side of Eq. (6.57)

$$\begin{aligned} 2Q_A &= \sum_{\mu\in A} 2(\mathbf{DS})_{\mu\mu} = \sum_{\mu\in A} [(\mathbf{DS})^2]_{\mu\mu} = \sum_{\mu\in A}\sum_{\nu}(\mathbf{DS})_{\mu\nu}(\mathbf{DS})_{\nu\mu} \\ &= \sum_{B}\sum_{\mu\in A}\sum_{\nu\in B}(\mathbf{DS})_{\mu\nu}(\mathbf{DS})_{\nu\mu} = \sum_{\substack{B\\ B\neq A}} B_{AB} + \sum_{\mu,\nu\in A}(\mathbf{DS})_{\mu\nu}(\mathbf{DS})_{\nu\mu} \,, \end{aligned} \tag{6.58}$$

where the definition (6.18) valid in the closed shell (singlet) case has been utilized. Regrouping the terms we see that in the RHF case (doubly filled orbitals) one indeed has

$$V_A^{\mathrm{RHF}} = \sum_{\substack{B\\ B\neq A}} B_{AB} \quad . \tag{6.59}$$

In the general—UHF, ROHF or correlated—case equality (6.59) does not hold, and one can define [24,25,63,64] the *free valence* F_A of the atom as the difference

$$F_A = V_A - \sum_{\substack{B\\ B\neq A}} B_{AB} \quad . \tag{6.60}$$

It follows from this definition that $F_A \equiv 0$ for all systems described with a determinant built up of doubly filled orbitals. In the UHF case the free valence index F_A can be expressed through the spin-density matrix $\mathbf{P}^s = \mathbf{P}^\alpha - \mathbf{P}^\beta$ [25,63,64]. To see this, we again start from the term $2Q_A$ on the right-hand side of Eq. (6.57) and use the idempotency relationships $(\mathbf{P}^\sigma\mathbf{S})^2 = \mathbf{P}^\sigma\mathbf{S}$, ($\sigma = \alpha$ *or* β). We have:

$$
\begin{aligned}
2Q_A &= 2\sum_{\mu\in A}\left[(\mathbf{P}^\alpha\mathbf{S})_{\mu\mu}) + (\mathbf{P}^\beta\mathbf{S})_{\mu\mu}\right] \\
&= 2\sum_{\mu\in A}\sum_{\nu}\left[(\mathbf{P}^\alpha\mathbf{S})_{\mu\nu}(\mathbf{P}^\alpha\mathbf{S})_{\nu\mu} + (\mathbf{P}^\beta\mathbf{S})_{\mu\nu}(\mathbf{P}^\beta\mathbf{S})_{\nu\mu}\right] \\
&= 2\sum_{\mu,\nu\in A}\left[(\mathbf{P}^\alpha\mathbf{S})_{\mu\nu}(\mathbf{P}^\alpha\mathbf{S})_{\nu\mu} + (\mathbf{P}^\beta\mathbf{S})_{\mu\nu}(\mathbf{P}^\beta\mathbf{S})_{\nu\mu}\right] + \sum_{\substack{B\\ B\neq A}} B_{AB} \\
&= \sum_{\mu,\nu\in A}\left[(\mathbf{DS})_{\mu\nu}(\mathbf{DS})_{\nu\mu} + (\mathbf{P}^s\mathbf{S})_{\mu\nu}(\mathbf{P}^s\mathbf{S})_{\nu\mu}\right] + \sum_{\substack{B\\ B\neq A}} B_{AB} \quad ,
\end{aligned}
\tag{6.61}
$$

substituting which into Eq. (6.60) and using definition (6.57) of the total valence, one gets that in the single determinant (ROHF or UHF) case [25,63,64]

$$
F_A = \sum_{\mu,\nu\in A}(\mathbf{P}^s\mathbf{S})_{\mu\nu}(\mathbf{P}^s\mathbf{S})_{\nu\mu} \quad . \tag{6.62}
$$

This result indicates that the free valence F_A measures the radical character of the atom in question. This is indeed observed if one considers an organic radical treated at the UHF level: one obtains a value of F_A close to 1 at the radical center and rather small values elsewhere. (Note that F_A is quadratic in the spin density.) Although F_A is sometimes called the number of uncompensated spins on the atom, that is not quite true, at least in the sense that the sum of the free valences does not equal the number of unpaired electrons in the system. In the simplest case, when there is a single unpaired electron besides closed shells, and it delocalizes only to a single orbital of each atom, then the free valence of each atom is equal to the square of its spin population, thus—as the spin populations sum to 1—the sum of the *square roots* of the free valences is also 1.* In general, especially in the UHF case the sum of square roots of the free valences usually slightly exceeds one. (We shall discuss the "number of effectively free electrons" in the next section.)

Table 6.7 displays the valence values for the prototype molecules for which the bond orders have been shown in Table 6.2 (For closed shell diatomic molecules the valence of both atoms equals the bond order, so they are not included in the table. Also, the three hydrogens in the CH_3 groups in the propane and ethanol molecules are not strictly equivalent, but the difference of their valences is so small that they are not shown separately.)

*The relation between the spin population and spin density matrix is by far not that simple in the general case. It may even happen that each atom of a molecule exhibits zero UHF spin population while the spin density matrix does not vanish—e.g., spins α and β occupy different p-orbitals on every atom.

Table 6.7
Valences in some prototype molecules calculated by using STO-3G, 6-31G**, and cc-pVTZ basis sets at the equilibrium bond distances

Molecule	Atom	Valence		
		STO-3G	6-31G**	cc-pVTZ
Methane	C	3.965	3.910	3.909
	H	0.996	0.953	0.980
Ethane	C	3.968	3.862	3.892
	H	0.997	0.949	0.978
Ethylene	C	3.976	3.893	3.822
	H	0.996	0.958	0.972
Acetylene	C	3.986	4.115	3.725
	H	0.988	0.930	0.953
Propane	C_1	3.967	3.851	3.891
	C_2	3.967	3.825	3.860
	H (CH_3)	0.997	0.951	0.980
	H (CH_2)	0.997	0.946	0.977
trans-Butadiene	C_1	3.975	3.899	3.835
	C_2	3.977	3.878	3.764
	H (C_1)	0.996	0.960	0.978
		0.996	0.959	0.976
	H (C_2)	0.996	0.957	0.976
Benzene	C	3.979	3.922	3.800
	H	0.996	0.954	0.977
Methyl radical	C total	3.960	3.905	3.871
	free	1.031	1.015	0.957
	H total	0.997	0.961	0.981
	free	0.009	0.009	0.006
Water	O	1.928	1.768	1.996
	H	0.973	0.883	1.012
HCN	C	3.957	3.824	4.008
	N	2.999	2.948	3.101
	H	0.978	0.907	0.948
CN radical	C total	3.897	3.846	4.060
	free	1.735	1.339	1.279
	N total	3.032	2.935	3.120
	free	0.871	0.428	0.339
Ethanol	C_1	3.933	3.761	3.934
	C_2	3.965	3.851	3.885
	O	1.990	1.729	1.999
	H (CH_2)	0.998	0.943	0.969
	H (CH_3)	0.996	0.951	0.985
	H (O)	0.971	0.884	1.027

The free valence index is a more informative quantity in the correlated case: it measures the extent to which the electronic structure around the given atom deviates from a situation with doubly occupied orbitals. And such a deviation does not necessarily mean the presence of any spin density; a significant free valence may occur even inside of a global singlet system for which the spin density identically vanishes everywhere.

One can obtain the explicit expression of the free valence index in the correlated case in a manner similar to the single determinant one. It follows from Eq. (6.49) that

$$2Q_A = \sum_{\mu\in A}\sum_{\nu}\left[(\mathbf{DS})_{\mu\nu}(\mathbf{DS})_{\nu\mu} + (\mathbf{RS})_{\mu\nu}(\mathbf{RS})_{\nu\mu}\right] \quad , \tag{6.63}$$

substituting which into the definition of valence (6.57) and then using also the definition (6.50) of the bond order, one gets for the right-hand side of Eq. (6.60)

$$F_A = \sum_{\mu,\nu\in A} (\mathbf{RS})_{\mu\nu}(\mathbf{RS})_{\nu\mu} \quad . \tag{6.64}$$

This shows that matrix **R** has a role analogous to the spin-density matrix in this case, too. In practice the use of Eq. (6.60) suffices in both single determinant and correlated cases.

The method of defining the bond order and valence indices described above is devoted to both single determinantal and correlated wave functions. It is *very chemical*, and is applicable not only near the equilibrium distances, but is able to describe the whole process of bond formation/dissociation. This is illustrated well in Figure 6.2 which displays some results for the ethylene molecule dissociating into two triplet methylenes, as calculated with a (4,4) CAS wave function by using cc-pVTZ basis set. One may see that the C–C bond order which is nearly two at the equilibrium distance, gradually decreases and tends to zero at the large distances—as it should. Simultaneous with this, there appears a free valence on the carbon, tending to a limit close to two at the large distances, in agreement with the fact that there are two unpaired electrons in the triplet methylene. The sum of the C–C bond order and of the carbon free valence is almost constant, thus the carbon atom remains practically four-valent during the whole dissociation.* (The C–H bond order stays practically constant at a value rather close to one.)

It is to be stressed that the DFT method is used to take into account the electron correlation in a heuristic or semiempirical manner, but actually it uses a

*It may be mentioned in this connection that the ground state of methylene is the triplet; contrary to the CAS scheme, the RHF method is only able to describe dissociation of ethylene into two singlet methylenes and results in *divalent* carbons with no free valences.

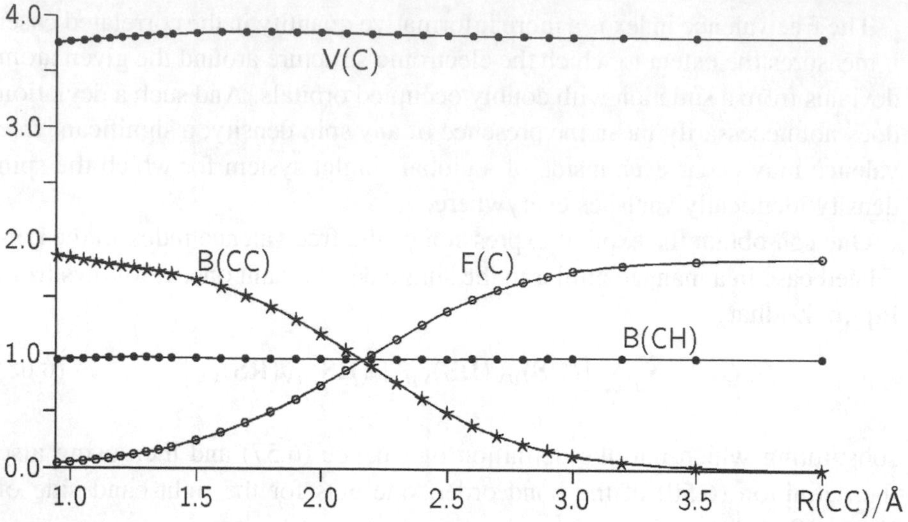

Figure 6.2

C–C and C–H bond orders, and total and free valences of the carbon for the dissociation of the ethylene molecule into two triplet methylenes, treated at the CAS(4,4) level of theory by using cc-pVTZ basis set. [66] with permission from Elsevier.

single determinant Kohn–Sham wave function. Therefore, the calculation of the bond order and valence indices—and also that of effective AOs, local spins, etc.—does not differ in the DFT case from that used in the Hartree–Fock case. At the same time, the energy decompositions, like those discussed in Chapter 8, in the DFT case met conceptual difficulties that we could not solve in the Hilbert-space analysis and were very involved in the 3D one, so we shall not enter into their considerations.

6.11 Hydrogen bonded systems

The hydrogen bonded systems X–H…Y may also be considered a special case of the three-center four-electron systems discussed in Section 6.5. At first insight, two electrons form the two-center X–H bond and another two occupy a lone pair on atom Y. However, the lone pair electrons exhibit some delocalization on the antibonding σ^* orbital of X–H moiety that leads to some weakening of the X–H bond, but also to the appearance of a partial H…Y bond. It may be worth looking at the problem from the point of view of bond order and valence indices.

We restrict ourselves to the consideration of one (effective) atomic orbital on the hydrogen atom. (The results of Chapter 5 indicate that this is a reasonable approximation.) Then the valence (6.57) of the hydrogen atom can be expressed—in agreement with Wiberg's formula (6.54)—as

$$V_H = 2Q_H - Q_H^2 , \tag{6.65}$$

where V_H is the valence of the hydrogen atom and Q_H is its Mulliken's gross atomic population. We introduce the resulting charge δ_H of the hydrogen, by taking into account the nuclear charge $Z_H = 1$, as

$$\delta = 1 - Q_H , \tag{6.66}$$

and express V_H in terms of δ_H. We get

$$V_H = 1 - \delta^2 . \tag{6.67}$$

This result indicates that the valence of the hydrogen atom depends only *quadratically* on the resulting charge on it; say for the relatively wide range of $\delta = [-0.2, 0.2]$, the valence V_H varies in the narrow range of $[0.96, 1]$.

When the hydrogen bond is formed, there appears a significant bond order also with the oxygen atoms of the partner molecule. During that, the valence of the hydrogen atoms stays close to unity according to the above discussion. At the same time the valence of the atom—if a closed shell wave function is used—is equal to the sum of the bond orders of that atom, c.f. Eq. (6.59). Therefore the formation of the hydrogen bond that is connected with the appearance of some bond order with the partner proton acceptor atom necessarily leads to an appropriate decrease of the intramolecular X–H bond order, indicating a weakening of that bond. This weakening can be observed in practice both as the increase of the X–H bond length and the reduction of the X–H stretching frequency.

6.12 Partial valences and some remarks on hypervalency

As discussed in Section 6.10, in the closed shell case the valence of atom A is given by the formula

$$V_A = 2Q_A - \sum_{\mu,\nu\in A} (\mathbf{DS})_{\mu\nu}(\mathbf{DS})_{\nu\mu} = 2\sum_{\mu\in A} (\mathbf{DS})_{\mu\mu} - \sum_{\mu,\nu\in A} (\mathbf{DS})_{\mu\nu}(\mathbf{DS})_{\nu\mu} \quad . \tag{6.68}$$

This can be collected as a sum over the basis orbitals (index μ) of the atom:

$$V_A = \sum_{\mu\in A} \left\{2(\mathbf{DS})_{\mu\mu} - \sum_{\nu\in A} (\mathbf{DS})_{\mu\nu}(\mathbf{DS})_{\nu\mu}\right\} \quad . \tag{6.69}$$

This grouping can be justified by observing that each individual term in the braces is nothing else than the sum of the *partial bond orders* $(\mathbf{DS})_{\mu\nu}(\mathbf{DS})_{\nu\mu}$ formed by orbital μ with all the orbitals ν belonging to *other atoms*:

$$V_A = \sum_{\mu\in A} \sum_{\substack{B \\ A\neq B}} \sum_{\nu\in B} (\mathbf{DS})_{\mu\nu}(\mathbf{DS})_{\nu\mu} \quad . \tag{6.70}$$

That equality can easily be proved by substituting into Eq. (6.69) the identity $2(\mathbf{DS})_{\mu\mu} = \sum_{\nu} (\mathbf{DS})_{\mu\nu}(\mathbf{DS})_{\nu\mu}$, following from the idempotency relationship $2\mathbf{DS} = \mathbf{DS}^2$, and separating the intraatomic terms that cancel with the second sum in Eq. (6.69).

Thus the valence of an atom V_A may be considered as a sum of *partial valences* corresponding to the individual atomic orbitals, given by the terms of the sums in Eqs. (6.69) or (6.70). However, the different basis orbitals (except the s-type ones) mix over when the molecule is rotated as a whole, so it is rather a meaning to consider partial valences belonging to a given azimuthal number l, i.e., those corresponding to the s, p, d, etc., type of orbitals:

$$V_A^{(l)} = \sum_{\mu\in A}^{(l)} \left\{2(\mathbf{DS})_{\mu\mu} - \sum_{\nu\in A} (\mathbf{DS})_{\mu\nu}(\mathbf{DS})_{\nu\mu}\right\} = \sum_{\mu\in A}^{(l)} \sum_{\substack{B \\ A\neq B}} \sum_{\nu\in B} (\mathbf{DS})_{\mu\nu}(\mathbf{DS})_{\nu\mu} \quad . \tag{6.71}$$

Table 6.8 displays the partial valences of different atoms in the CH_3SO_2Cl molecule already considered in Section 5.2, calculated by using cc-pVTZ basis set. The results agree very well with the chemical expectations. Thus

Table 6.8
Partial valences of the different atoms in the CH_3SO_2Cl molecule calculated by using cc-pVTZ basis set

Atom	Valence contribution				
	s	*p*	*d*	*f*	Total
C	0.887	2.888	0.068	0.004	3.847
H	0.932	0.052	0.004	0.000	0.987
H	0.936	0.055	0.004	0.000	0.996
H	0.932	0.052	0.004	0.000	0.987
S	0.880	2.961	1.777	0.199	5.816
O	0.264	1.776	0.029	0.002	2.070
O	0.264	1.776	0.029	0.002	2.070
Cl	0.040	0.961	0.046	0.004	1.051

*[53] with permission from Wiley.

carbon practically has a valence four, one of which is given by the *s*-orbitals, three by the *p*-orbitals. Oxygen and chlorine are divalent and monovalent, respectively, and their valence is basically due to the *p*-type orbitals, while the valence of the monovalent hydrogens is to be attributed to their *s*-orbitals. Remarkable are the results obtained for sulfur: its *s*- and *p*-valences are similar to those of carbon, but there is a valence close to two which is due to the *d*-orbitals—and there is even a small, but non-negligible, *f*-contribution. As a result, the sulfur has an overall valence approaching six—as many as the number of the valence bars on the classical structural formula.

That effect of the *d*-orbitals can be better understood, if one takes into account that the quantities (6.54)

$$b_\mu = 2q_\mu - q_\mu^2 = 2(\mathbf{DS})_{\mu\mu} - (\mathbf{DS})_{\mu\mu}^2 \tag{6.72}$$

are nonlinear functions of the gross orbital populations $q_\mu = (\mathbf{DS})_{\mu\mu}$: five d−orbitals having an average gross orbital population $q_\mu = 0.2$ each (which is roughly the case for the sulfur in CH_3SO_2Cl) will give diagonal contributions to the atomic valence $5(2 \cdot 0.2 - 0.2^2) = 1.8$; the resulting partial *d*-orbital valence in Table 6.8 is not far from this value.

These results, together with those discussed in Section 5.2, indicate that one can find a compromise between two different descriptions of the hypervalent systems—the old one assuming the *d*-orbitals to behave as valence ones, and the newer stressing the ionic character of these molecules. According to our analysis, these atoms have very strong ionic character, but the back-donation to their d-orbitals (and even possibly f-orbitals) is so significant that these orbitals are able to account for the atoms approaching the classical formal valences four or six.

Similar, but significantly weaker effect can be observed, e.g., in the case of carbon atoms that are positive because they are connected to electronegative oxygen(s)—like the central carbon in acetone or the carboxylic one in glycine. These atoms also exhibit a significant (~0.4–0.5) d-valence, due to which their overall valence slightly exceeds four—while in most cases the Hilbert-space valences of the atoms are somewhat lower than their ideal integer values.

7

Open-shell systems and local spins

7.1 Effective number of unpaired electrons

As known, the spinless density matrix of a closed shell (singlet) single determinant wave function has the peculiar idempotency property

$$(\mathbf{DS})^2 = 2\mathbf{DS} \quad , \tag{7.1}$$

which has also been utilized several times above. If the wave function is not a closed shell singlet or if electron correlation has been taken into account, then this relation is not satisfied, and the number

$$\Delta = Tr[2\mathbf{DS} - (\mathbf{DS})^2] = \sum_{\mu}[2\mathbf{DS} - (\mathbf{DS})^2]_{\mu\mu} \quad , \tag{7.2}$$

gives a measure of the deviation from the closed shell situation in the molecule, and it can be called the number of effectively unpaired electrons [72,81]. An important property of it is that it is always non-negative, which follows from the fact that matrix $\mathbf{D}$ can be expressed in terms of the coefficient vectors $\mathbf{c}_i$ of the expansion the natural orbitals $\varphi_i = \sum_\mu c^i_\mu \chi_\mu$ in terms of the basis orbitals χ_μ, and their occupation numbers n_i as

$$\mathbf{D} = \sum_i n^i \mathbf{c}_i \mathbf{c}_i^\dagger \quad . \tag{7.3}$$

By substituting Eq. (7.3) into (7.2) and taking into account the orthonormalization $\mathbf{c}_i^\dagger \mathbf{S} \mathbf{c}_j = \delta_{ij}$ of the natural orbitals, and the relationship $Tr(\mathbf{c}_i \mathbf{c}_i^\dagger \mathbf{S}) = Tr(\mathbf{c}_i^\dagger \mathbf{S} \mathbf{c}_i) = 1$, we get

$$\Delta = \sum_i (2n_i - n_i^2) = \sum_i n_i(2 - n_i) \quad . \tag{7.4}$$

As the occupation numbers are in the interval $0 \leq n_i \leq 2$, neither of the components of this sum can be negative.

By writing $\mathbf{D} = \mathbf{P}^\alpha + \mathbf{P}^\beta$, as well as adding and subtracting $2[(\mathbf{P}^\alpha\mathbf{S})^2 + (\mathbf{P}^\beta\mathbf{S})^2]$, the matrix $2\mathbf{DS} - (\mathbf{DS})^2$ can be transformed as

$$\begin{aligned} 2\mathbf{DS} - (\mathbf{DS})^2 &= 2(\mathbf{P}^\alpha\mathbf{S} + \mathbf{P}^\beta\mathbf{S}) - (\mathbf{P}^\alpha\mathbf{S} + \mathbf{P}^\beta\mathbf{S})^2 \\ &= 2(\mathbf{P}^\alpha\mathbf{S} + \mathbf{P}^\beta\mathbf{S}) - (\mathbf{P}^\alpha\mathbf{S})^2 - \mathbf{P}^\alpha\mathbf{S}\mathbf{P}^\beta\mathbf{S} - \mathbf{P}^\beta\mathbf{S}\mathbf{P}^\alpha\mathbf{S} - (\mathbf{P}^\beta\mathbf{S})^2 \\ &= 2[\mathbf{P}^\alpha\mathbf{S} + \mathbf{P}^\beta\mathbf{S} - (\mathbf{P}^\alpha\mathbf{S})^2 - (\mathbf{P}^\beta\mathbf{S})^2] \\ &\qquad + (\mathbf{P}^\alpha\mathbf{S})^2 - \mathbf{P}^\alpha\mathbf{S}\mathbf{P}^\beta\mathbf{S} - \mathbf{P}^\beta\mathbf{S}\mathbf{P}^\alpha\mathbf{S} + (\mathbf{P}^\beta\mathbf{S})^2 \\ &= \mathbf{P}^s\mathbf{S}\mathbf{P}^s\mathbf{S} + 2[\mathbf{P}^\alpha\mathbf{S} - (\mathbf{P}^\alpha\mathbf{S})^2 + \mathbf{P}^\beta\mathbf{S} - (\mathbf{P}^\beta\mathbf{S})^2] \quad . \end{aligned} \tag{7.5}$$

In the last line the spin-density matrix $\mathbf{P}^s = \mathbf{P}^\alpha - \mathbf{P}^\beta$ has also been introduced. Equation (7.5) indicates that there are two types of contributions to the effective number of unpaired electrons: one is directly connected with the spin-density matrix, another with the deviation of the density matrices $\mathbf{P}^\alpha$ and $\mathbf{P}^\beta$ from the idempotency property characteristic for the single determinant wave functions. Obviously, the terms of the first type vanish for any pure singlet state for which $\mathbf{P}^s \equiv \mathbf{0}$, while those of the second type vanish whenever the wave function can be presented as a single determinant.

It is worth considering some special cases in some detail.

ROHF case with N_{open} unpaired electrons of α spin

In this case $\mathbf{D} = 2\mathbf{P}^c + \mathbf{P}^s$, where $\mathbf{P}^c$ corresponds to closed-shell part and $\mathbf{P}^s$ is the spin-density matrix corresponding to the open-shell electrons. As we are dealing with a single determinant wave function built up of orthonormalized orbitals, it can easily be seen that in this case the spin-density matrix itself is idempotent: $(\mathbf{P}^s\mathbf{S})^2 = \mathbf{P}^s\mathbf{S}$. Thus Eq. (7.5) reduces to $2\mathbf{DS} - (\mathbf{DS})^2 = \mathbf{P}^s\mathbf{S}$, and we get

$$\Delta = Tr(\mathbf{P}^s\mathbf{S}) = \sum_A \sum_{\mu\in A} (\mathbf{P}^s\mathbf{S})_{\mu\mu} = \sum_A p^s_A = N_{open} \quad , \tag{7.6}$$

where

$$p^s_A = \sum_{\mu\in A} (\mathbf{P}^s\mathbf{S})_{\mu\mu} \quad , \tag{7.7}$$

is the gross spin population of atom A. The fact that in this simplest case Δ equals the number of unpaired electrons justifies its name.

Case of a UHF single determinant wave function

In this case Eq. (7.5) gives

$$2\mathbf{DS} - (\mathbf{DS})^2 = (\mathbf{P}^s\mathbf{S})^2 \quad . \tag{7.8}$$

Thus

$$
\begin{aligned}
\Delta = Tr[2\mathbf{DS} - (\mathbf{DS})^2] &= \sum_{\mu}[(\mathbf{P}^s\mathbf{S})^2]_{\mu\mu} = \sum_{\mu,\nu}(\mathbf{P}^s\mathbf{S})_{\mu\nu}(\mathbf{P}^s\mathbf{S})_{\nu\mu} \\
&= \sum_{A}\sum_{\mu,\nu\in A}(\mathbf{P}^s\mathbf{S})_{\mu\nu}(\mathbf{P}^s\mathbf{S})_{\nu\mu} + \sum_{\substack{A,B \\ A\neq B}}\sum_{\mu\in A}\sum_{\nu\in B}(\mathbf{P}^s\mathbf{S})_{\mu\nu}(\mathbf{P}^s\mathbf{S})_{\nu\mu} \qquad (7.9) \\
&= \sum_{A} F_A + \sum_{\substack{A,B \\ A\neq B}}\sum_{\mu\in A}\sum_{\nu\in B}(\mathbf{P}^s\mathbf{S})_{\mu\nu}(\mathbf{P}^s\mathbf{S})_{\nu\mu} \quad ,
\end{aligned}
$$

owing to Eq. (6.62) relevant in this case.

Of course, the ROHF result Eq. (7.6) can also be transformed to this form. In fact, in that case $\mathbf{P}^s\mathbf{S} = (\mathbf{P}^s\mathbf{S})^2$, thus the right-hand side of Eq. (7.6) can be rewritten to that of Eq. (7.9).

Singlet correlated case

In this case one can express Δ through matrix **RS** defined in Eq. (6.49):

$$
\begin{aligned}
\Delta &= \sum_{\mu,\nu}(\mathbf{RS})_{\mu\nu}(\mathbf{RS})_{\nu\mu} \\
&= \sum_{A}\sum_{\mu,\nu\in A}(\mathbf{RS})_{\mu\nu}(\mathbf{RS})_{\nu\mu} + \sum_{\substack{A,B \\ A\neq B}}\sum_{\mu\in A}\sum_{\nu\in B}(\mathbf{RS})_{\mu\nu}(\mathbf{RS})_{\nu\mu} \qquad (7.10) \\
&= \sum_{A} F_A + \sum_{\substack{A,B \\ A\neq B}}\sum_{\mu\in A}\sum_{\nu\in B}(\mathbf{RS})_{\mu\nu}(\mathbf{RS})_{\nu\mu}
\end{aligned}
$$

which puts the concept of the free valence into a new perspective. (Matrix **RS** again plays a role analogous to the spin-density matrix of an open-shell single determinant.)

The relationship between Δ and the cumulant of the second-order density matrix in the general correlated case will be considered in the next section.

Spatial distribution of the "effectively unpaired" electrons

Δ can also be considered as the spatial integral of a density-like quantity

$$
\Delta = \sum_i \int n_i(2-n_i)\varphi_i^*(\vec{r})\varphi_i(\vec{r})dv = \int \varrho_u(\vec{r})dv \quad . \qquad (7.11)
$$

According to its definition, $\varrho_u(\vec{r}) \geq 0$ everywhere. It can also written as [81]

$$
\varrho_u(\vec{r}) = 2\varrho(\vec{r}) - \int \varrho_1(\vec{r};\vec{r}')\varrho_1(\vec{r}';\vec{r})dv' \quad , \qquad (7.12)
$$

with $\varrho_1(\vec{r};\vec{r}')$ being the spinless first-order density matrix defined in Eq. (4.2). In fact, the expansion of $\varrho_1(\vec{r};\vec{r}')$ in terms of the natural orbitals is

$$\varrho_1(\vec{r};\vec{r}') = \sum_i n_i \varphi_i^*(\vec{r}')\varphi_i(\vec{r}) \quad , \tag{7.13}$$

and equality (7.12) follows from the orthonormalization of the natural orbitals. Furthermore, $\varrho(\vec{r}) \equiv \varrho_1(\vec{r};\vec{r})$ is the electron-density—diagonal term (4.1) of the spinless density matrix.

Thus $\varrho_u(\vec{r})$ may be considered the three-dimensional density of the open-shell electrons. By using the definition Eq. (4.2), one obtains its basis set expansion as

$$\varrho_u(\vec{r}) = \sum_{\mu,\nu}(2\mathbf{D}-\mathbf{DSD})_{\mu\nu}\chi_\nu^*(\vec{r})\chi_\mu(\vec{r}) = 2\varrho(\vec{r}) - \sum_{\mu,\nu}(\mathbf{DSD})_{\mu\nu}\chi_\nu^*(\vec{r})\chi_\mu(\vec{r}) \quad , \tag{7.14}$$

where the expression (3.16) of the electron density $\varrho(\vec{r})$ has been used. It is easy to see that in the closed shell RHF case $\varrho_u(\vec{r})$ vanishes in every point of space, as the idempotency relationship (7.1) can also be formulated as $2\mathbf{D} = \mathbf{DSD}$. Also, in the ROHF case $\varrho_u(\vec{r})$ simply equals the spin density.

7.2 Effective number of unpaired electrons and the cumulant

The effective number of unpaired electrons Δ is expressed through the first-order density matrix and its spatial distribution $\varrho_u(\vec{r})$ depends only on a single spatial coordinate. Nonetheless, it is clear that these quantities are not of genuine one-electron nature, as they reflect whether the different electrons are effectively paired or not. Accordingly, it is not surprising that one can find an intimate connection with the second-order density matrix [82]. As for single determinant wave functions the second-order density matrix can be expressed through the first-order one, this becomes of direct importance for correlated wave functions.

For the sake of simplicity, in this section we shall use some *orthonormalized* set of spin-orbitals $\{\psi_i\}$ and the creation and annihilation operators $\hat{\psi}_i^+$ and $\hat{\psi}_i^-$ corresponding to them. As known [83] (also see Appendix A.1), the matrix element P_{ij} of the first-order density matrix and a hypermatrix-element Γ_{lkij} of the second-order density matrix can be expressed as the expectation values of the operator strings $\hat{\psi}_j^+\hat{\psi}_i^-$ and $\hat{\psi}_i^+\hat{\psi}_j^+\hat{\psi}_k^-\hat{\psi}_l^-$, respectively:

$$P_{ij} = \langle \hat{\psi}_j^+\hat{\psi}_i^- \rangle \, ; \qquad \Gamma_{lkij} = \langle \hat{\psi}_i^+\hat{\psi}_j^+\hat{\psi}_k^-\hat{\psi}_l^- \rangle \quad . \tag{7.15}$$

Actually we will have to specify explicitly also the spin σ corresponding to these operators, and write $\hat{\psi}_i^{\sigma+}$ and $\hat{\psi}_i^{\sigma-}$. It will be assumed that the spin-orbitals ψ_i^α and ψ_i^β have identical spatial parts.

The *cumulant* $\mathbf{\Lambda}$ of the second-order density matrix $\mathbf{\Gamma}$ is that component of it which cannot be expressed through the elements of the first-order density matrix $\mathbf{P}^\sigma$ in a manner characteristic for the single determinant wave functions; it may be considered a correction term:

$$\begin{aligned}\Gamma_{l^\sigma k^{\sigma'} i^{\sigma''} j^{\sigma'''}} &= \langle \hat{\psi}_i^{\sigma''+} \hat{\psi}_j^{\sigma'''+} \hat{\psi}_k^{\sigma'-} \hat{\psi}_l^{\sigma-} \rangle \\ &= \delta_{\sigma\sigma''}\delta_{\sigma'\sigma'''} P_{li}^{\sigma} P_{kj}^{\sigma'} - \delta_{\sigma\sigma'''}\delta_{\sigma'\sigma''} P_{lj}^{\sigma} P_{ki}^{\sigma'} + \Lambda_{l^\sigma k^{\sigma'} i^{\sigma''} j^{\sigma'''}} \quad .\end{aligned} \tag{7.16}$$

Accordingly, the cumulant vanishes identically for any wave function which can be written down as a single determinant.

We define also the "spin-less cumulant" Λ_{lkij} as the sum of the four respective spin-dependent ones:

$$\Lambda_{lkij} = \Lambda_{l^\alpha k^\alpha i^\alpha j^\alpha} + \Lambda_{l^\alpha k^\beta i^\alpha j^\beta} + \Lambda_{l^\beta k^\alpha i^\beta j^\alpha} + \Lambda_{l^\beta k^\beta i^\beta j^\beta} \tag{7.17}$$

We shall consider the sum of expectation values of the following four operator-strings, and expand it by using Eqs. (7.16) and (7.17):

$$\begin{aligned}&\langle \hat{\psi}_i^{\alpha+} \hat{\psi}_j^{\alpha+} \hat{\psi}_j^{\alpha} \ \hat{\psi}_l^{\alpha-} \rangle + \langle \hat{\psi}_i^{\alpha+} \hat{\psi}_j^{\beta+} \hat{\psi}_j^{\beta-} \hat{\psi}_l^{\alpha-} \rangle \\ &\qquad + \langle \hat{\psi}_i^{\beta+} \hat{\psi}_j^{\alpha+} \hat{\psi}_j^{\alpha-} \hat{\psi}_l^{\beta-} \rangle + \langle \hat{\psi}_i^{\beta+} \hat{\psi}_j^{\beta+} \hat{\psi}_j^{\beta-} \hat{\psi}_l^{\beta-} \rangle \\ &= P_{li}^\alpha P_{jj}^\alpha - P_{lj}^\alpha P_{ji}^\alpha + P_{li}^\alpha P_{jj}^\beta + P_{li}^\beta P_{jj}^\alpha + P_{li}^\beta P_{jj}^\beta - P_{lj}^\beta P_{ji}^\beta + \Lambda_{ljij} \\ &= D_{li} D_{jj} - \tfrac{1}{2}\left[D_{lj} D_{ji} + P_{lj}^s P_{ji}^s\right] + \Lambda_{ljij}\end{aligned} \tag{7.18}$$

where the last equality has been obtained after adding and subtracting $P_{lj}^\alpha P_{ji}^\beta + P_{lj}^\beta P_{ji}^\alpha$.

Now, we sum over the repeated subscript j. On the right-hand side the sum of the terms D_{jj} obviously equals N, the number of electrons. The sum of the expectation values on the left-hand side can be transformed by interchanging first the two annihilator operators and then the creation and annihilation operator in the middle of each string, utilizing the Fermion anticommutation rules to get

$$\begin{aligned}-\sum_j \Big[&\langle \hat{\psi}_i^{\alpha+} (\delta_{jl} - \hat{\psi}_l^{\alpha-} \hat{\psi}_j^{\alpha+}) \hat{\psi}_j^{\alpha-} \rangle - \langle \hat{\psi}_i^{\alpha+} \hat{\psi}_l^{\alpha-} \hat{\psi}_j^{\beta+} \hat{\psi}_j^{\beta-} \rangle \\ &- \langle \hat{\psi}_i^{\beta+} \hat{\psi}_l^{\beta-} \hat{\psi}_j^{\alpha+} \hat{\psi}_j^{\alpha-} \rangle + \langle \hat{\psi}_i^{\beta+} (\delta_{jl} - \hat{\psi}_l^{\beta-} \hat{\psi}_j^{\beta+}) \hat{\psi}_j^{\beta-} \rangle \Big]\end{aligned}$$

$$
\begin{aligned}
&= -\langle\hat{\psi}_i^{\alpha+}\hat{\psi}_l^{\alpha-}\rangle - \langle\hat{\psi}_i^{\beta+}\hat{\psi}_l^{\beta-}\rangle \\
&\quad + \langle\hat{\psi}_i^{\alpha+}\hat{\psi}_l^{\alpha-}(\hat{N}_\alpha + \hat{N}_\beta)\rangle + \langle\hat{\psi}_i^{\beta+}\hat{\psi}_l^{\beta-}(\hat{N}_\alpha + \hat{N}_\beta)\rangle \\
&= (P_{li}^\alpha + P_{li}^\beta)(N-1) = D_{li}(N-1)
\end{aligned} \tag{7.19}
$$

where we have utilized that

$$
\hat{N}_\sigma = \sum_j \hat{\psi}_j^{\sigma+}\hat{\psi}_j^{\sigma-} \tag{7.20}
$$

is the operator of the number of electrons of spin σ. As we consider only wave functions with a fixed number of electrons, which are eigenvectors of the operator $\hat{N} = \hat{N}_\alpha + \hat{N}_\beta$ with the eigenvalue N, we could replace $\hat{N}_\alpha + \hat{N}_\beta$ in the expectation values by a simple multiplication with N. The terms $D_{li}N$ on the two sides cancel and we are left with

$$
-\tfrac{1}{2}\sum_j \left[D_{lj}D_{ji} + P_{lj}^s P_{ji}^s\right] + \sum_j \Lambda_{ljij} = -D_{li} \tag{7.21}
$$

which can be rewritten as

$$
-\tfrac{1}{2}(2\mathbf{D} - \mathbf{D}^2)_{li} = \sum_j \left[\Lambda_{ljij} - \tfrac{1}{2}P_{lj}^s P_{ji}^s\right] \tag{7.22}
$$

Comparison with Eq. (7.14) shows that the left-hand side contains the elements of the matrix $2\mathbf{D} - \mathbf{D}^2$, giving the expansion coefficients of $\varrho_u(\vec{r})$ in the present orthonormalized basis for which matrix $\mathbf{S}$ is a unit matrix. Taking the trace in Eq. (7.22), we get

$$
\Delta = -2\sum_{i,j} \left[\Lambda_{ijij} - \tfrac{1}{2}P_{ij}^s P_{ji}^s\right] \tag{7.23}
$$

In the single determinant case the cumulant Λ_{ljij} vanishes, and equations (7.22) and (7.23) essentially reduce to (7.8) and the first line of (7.9), respectively.

A remarkable aspect of these equations is that their left-hand sides depend only on the spin-less first-order density matrix. That means that for an open-shell correlated case they are independent of the actual S_z projection of the wave function, i.e., are identical for the different components of a multiplet. Therefore the combination of Λ_{ijij} and of the product of the spin-density matrix elements on the right-hand side must be S_z-independent, too. For that reason Alcoba et al. call this combination the "spin-free cumulant" [82]; the use of a term like "spin-projection independent" would be perhaps more fortunate, as "spin-free" can easily be confused with "spin-less."

Of course, if one wishes to express Δ in terms of non-orthogonal basis functions, then the elements of the overlap matrix appear not only for the spin-density matrix but also for the cumulant—both being quantities relevant to the overlapping basis. Equation (7.23] then becomes

$$\Delta = -2\sum_{\mu,\nu}\left[\sum_{\rho,\tau}\Lambda_{\mu\nu\rho\tau}S_{\rho\mu}S_{\tau\nu} - \tfrac{1}{2}(\mathbf{P}^s\mathbf{S})_{\mu\nu}(\mathbf{P}^s\mathbf{S})_{\nu\mu}\right] . \tag{7.24}$$

7.3 Local spins

The effective number of unpaired electrons Δ discussed in the previous section gives an insight to what extent the given molecule is an open-shell one. The free valences F_A representing the one-center contribution to Δ and the spatial density of the unpaired electrons $\varrho_u(\vec{r})$ permit to assign this open-shell character to the individual atoms and to different parts of the 3-dimensional space, respectively. As an electron is a particle with spin, it is natural to assume that where there are unpaired electrons, there should be uncompensated spins, too. Spin density, giving the difference of the density of electrons with spins α and β, respectively

$$\varrho_s(\vec{r}) = \varrho_\alpha(\vec{r}) - \varrho_\beta(\vec{r}) = \sum_{\mu,\nu} P^s_{\mu\nu}\chi^*_\nu(\vec{r})\chi\mu(\vec{r}) \quad , \tag{7.25}$$

and its matrix counterpart $\mathbf{P}^s = \mathbf{P}^\alpha - \mathbf{P}^\beta$ give very important information about this, but that is only about the distribution of the z-projection of the overall spin and in a number of cases it is not sufficient to characterize the behavior of spins in the system. The most striking example is that of antiferromagnets which represent extended systems in the singlet state; therefore their spin density vanishes everywhere,* but their magnetic properties indicate the presence of two sub-lattices of oppositely oriented spins. The simplest of such systems are oxides or fluorides of transition metals with partially filled d-shells. The interaction between the d-shells of the neighboring centers is only indirect and is rather weak as compared to the relatively large one-center electron-electron repulsion; thus the electrons on these partly filled orbitals do not form conduction bands but give singly occupied orbitals with uncompensated local spins (so-called Mott insulators). These local spins are then coupled into a resulting singlet.

*To be precise, real antiferromagnets are characterized with an overall spin which is negligible macroscopically. Then the spin density can be neglected even at the microscopic scale.

Similar is the situation with several molecular systems, like bi- or polynuclear transition complexes or biradicals, in which the individual magnetic or radical centers often behave in the magnetic field resembling independent spins—despite the fact that the overall molecular state may be a singlet.

For instance, if we consider a fully dissociated H_2 molecule described by the simplest UHF method, then we get one spin up on atom A and one spin down on atom B—or *vice versa*. The spin populations $+1$ and -1 on the two atoms are giving a qualitative correct picture in that simple case, but one should keep in mind that these single determinants are not eigenfunctions of the total spin-square operator $\hat{S}^2$, i.e., are not pure spin states but represent a 1:1 mixture of a singlet and a triplet. Having pure spin states is important if we wish to connect the dissociated state with the overall potential curves of the system: we know that the singlet curve approaches the sum of the energies of two free hydrogen atoms from below, while the triplet curve from above: the singlet and triplet become degenerate only at a strictly infinite distance.

Forming the sum or difference of these single determinants one gets the pure singlet or the $S_z = 0$ component of the triplet (Heitler–London wave functions). For both the singlet and the $S_z = 0$ triplet the spin-density matrix is zero. But obviously, at the very large interatomic distances each hydrogen atom carries a single electron only, so its spin is not compensated: there is no other electron in the vicinity which would do that. There is no contradiction with the absence of the spin density: the z-projection of this spin is not defined (it can be either $+\frac{1}{2}$ or $-\frac{1}{2}$) but it is certain that this single electron has a spin-square equal to $\frac{3}{4}$.

It is obvious that all these four wave functions (the two UHF determinants differing in the interchange of spins and the two pure spin-eigenfunctions representing their sum and difference) lead in the asymptotic case to the same spinless first-order density matrix which is equal in terms of a minimal (or effective minimal—c.f. Section 5.1) basis to

$$\mathbf{D} = \begin{pmatrix} 1 & 0 \\ 0 & 1 \end{pmatrix} \quad . \tag{7.26}$$

(The matrices $\mathbf{P}^{\alpha}$ and $\mathbf{P}^{\beta}$ are not identical in all these cases.) The identical $\mathbf{D}$-matrices—and the absence of the two-center spin density or $\mathbf{R}$ matrix elements—also mean identical free valences F_A and identical effective numbers of unpaired electrons Δ. That is in agreement with the fact that all these wave functions essentially describe the same physical situation: the system consists of two free hydrogen atoms. In fact, at the truly infinite distance it should be nearly immaterial whether one forms a singlet or a triplet for the non-interacting atoms (or considers the mixture of these states). All these four states are characterized with a local spin-square of $\frac{3}{4}$ on both atoms, although the overall spin square has different expectation values $\langle \hat{S}^2 \rangle$ in their cases: 0 for the singlet, 2 for the triplet, and 1 (the average of the singlet and triplet values) for the two single determinants. As the local spin-squares are the same in all these cases $\langle \hat{S}^2 \rangle_A = \langle \hat{S}^2 \rangle_B = \frac{3}{4}$, the two-center contributions to $\langle \hat{S}^2 \rangle$ are different: $\langle \hat{S}^2 \rangle_{AB} = \langle \hat{S}^2 \rangle_{BA} = -\frac{3}{4}$ in the singlet case, $+\frac{1}{4}$ for the triplet and $-\frac{1}{4}$ (the average of the previous two) for the individual single determinants.

It follows from the above discussion that, in order to study the spin properties of a molecule other than spin density, one has to consider the decomposition of the expectation value of the total spin-square operator $\hat{S}^2$ into atomic and diatomic components. (One can measure simultaneously the square of the total spin and one of its projections, conventionally selected as the z-projection, so these are the two quantities we can deal with.) It is important to note once again that one can have non-zero local spins, i.e., atomic contributions to the total spin-square, even for a global singlet, for which $\langle \hat{S}^2 \rangle = 0$, because the two-center contributions can be of either sign—as we have seen above in the example of H_2.*

When doing the analysis of the spin-square, one encounters very sharply the problem that the partitioning of a single physical quantity into several components is usually not unique, and an infinite number of different decompositions may be introduced. Therefore, special care must be taken in order to get really meaningful results. Thus one should look for some additional requirements which permit to select one partition that may be considered privileged as giving physically reasonable results in all limiting cases.

The problem of non-uniqueness of decompositions is especially important in the case of decomposition of $\langle \hat{S}^2 \rangle$, because one encounters a very peculiar situation: the approach which would look to be the most straightforward one, leads to results which are unphysical in our opinion, as they do not permit us to distinguish between covalent molecules and antiferromagnetic systems.

*In this respect it may be noted that for a singlet state not only $\langle \hat{S}^2 \rangle = 0$, but the result of applying operator $\hat{S}$ to the wave function does already vanish. So seemingly we "do not have anything to decompose." But that resulting zero eigenvalue is the result of some factors compensating each other; therefore analyzing the $\langle \hat{S}^2 \rangle$ permits one to distinguish properly between the covalent and antiferromagnetic systems—both being singlets: for covalent systems the compensation takes place locally, while for the systems with antiferromagnetic coupling it is a consequence of some global effects pertinent to the molecule as a whole, leaving considerable local spins on the individual atoms.

Clark and Davidson [22, 84–86] have decomposed the total spin operator as the sum of atomic ones as

$$\hat{S} = \sum_A \hat{S}_A \quad , \tag{7.27}$$

with a choice $\hat{S}_A = \hat{S}\hat{P}_A$, where $\hat{P}_A$ is a projector of the atomic subspace of a Löwdin-orthogonalized basis, and calculated the different terms of the decomposition

$$\langle \hat{S}^2 \rangle = \sum_A \langle \hat{S}_A^2 \rangle + \sum_{\substack{A,B \\ A \neq B}} \langle \hat{S}_A \hat{S}_B \rangle \quad . \tag{7.28}$$

They got the result that the local spin-square of each atom of the H_2 molecule treated at the *closed shell* RHF level is $\frac{3}{8}$. In general, at the RHF level of theory the local spin-square of every atom is $\frac{3}{8}$ of its valence—which is then compensated by negative two-center components that are proportional to the respective bond orders. It should be noted that their derivation was correct and the conclusion remains valid if any reasonable definition of the atomic spin operators is used. Nonetheless, we feel that result is physically inappropriate: for instance a polyethylene chain would have a pretty large spin-square $\sim 4 \times \frac{3}{8} = 1.5$ on every carbon atom, i.e., it would be predicted to be an antiferromagnet.

In the formalism of Refs. 84–86, the diatomic spin components are obtained proportional to the respective bond orders taken with a negative sign, which provides the correct value of the overall $\langle \hat{S}^2 \rangle$. This means that not only those electronic spins are accounted for which are actually free, but also those which form singlet coupled pairs corresponding to the individual chemical bonds. Such method of counting spins does not satisfy us when we are interested in the radical nature of the system studied, and especially in the magnetic properties which may be related to the unpaired spins. It may be noted in this respect that the authors of Refs. 84–86 were mainly interested in the atomic spins appearing in the reformulation of the VB theory of electronic structure in terms of a formal magnetic Heisenberg Hamiltonian, rather than in the actual spin properties of the molecular systems considered.

From a physical point of view one is inclined to expect that systems treated in the closed shell RHF framework (doubly filled orbitals) are lacking any free spins, thus should not exhibit any non-zero local spins, nor any non-zero diatomic spin contributions. Thus in the case of closed shell covalent systems near their equilibrium configurations, for the description of which RHF is a good reference state, only insignificant local spins are expected even at the correlated level of the theory.

As has been noted above, one inevitably obtains counterintuitive results (significant local spins at the RHF level of theory) if one first performs any reasonable division of operator $\hat{S}$ into sum of atomic operators $\hat{S}_A$ and then calculates the expectation values $\langle \hat{S}_A^2 \rangle$ and $\langle \hat{S}_A \hat{S}_B \rangle$ of these atomic operators and of their scalar products, respectively. Instead of that, we shall start from the explicit expression of the expectation value $\langle \hat{S}^2 \rangle$ and decompose it into atomic and diatomic components only afterward. When doing this, we shall require the fulfillment of the following conditions:

(i) One should get *no spins whatever* for the covalent systems described by a *closed shell RHF wave function* using doubly filled orbitals.

(ii) For a properly dissociating wave function the *asymptotic values* of the atomic spins obtained for the atoms at large distances should coincide with the values pertinent to the respective *free atoms.* (As discussed above, in the example of the H_2 molecule, the decomposition of $\langle \hat{S}^2 \rangle$ can contain terms of different signs, so it is possible to obtain non-zero atomic spins even for a system which is in an overall singlet state.)

(iii) The decomposition formula should behave properly also in the special case of a system containing a *single electron.* As the property $\langle \hat{s}^2 \rangle$=3/4 is an *intrinsic* property of each electron, in the case of a single electron system one should consider the distribution of $\langle \hat{S}^2 \rangle$ as 3/4 times the distribution of the electron itself, so the decomposition of the spin-square should give results coinciding with those of the respective population analysis, multiplied by 3/4. (We should consider only gross populations, as the consideration of overlap populations would mean in the case of $\langle \hat{S}^2 \rangle$ assuming some spurious self-coupling of the spin of the single electron.) To fulfill this requirement is not that trivial as it could appear at the first insight, because of the peculiar character of the operator $\hat{S}^2$: it represents a sum of one-electron operators and two-electron ones, the treatment of which is somewhat intermixed when one takes care of the fulfillment of conditions (i) and (ii). So one should apply special measures to ensure the proper treatment of the systems containing a single electron—or a single unpaired electron outside the closed shells, which essentially represents the same problem.

(iv) For high-spin systems the values of the atomic and diatomic $\langle \hat{S}^2 \rangle$ components should not depend on the actual value of the S_z projection [87]. This means, that one should obtain, e.g., the same decomposition of $\langle \hat{S}^2 \rangle$ for the $S_z = 1$, $S_z = 0$ and $S_z = -1$ components of a triplet, and so on. The fulfillment of this requirement is possible if one can express the components through the spin-less first- and second-order density matrices.

7.3.1 Local spins for single determinant wave functions

We start* from known representation (given by Löwdin [90]) of the operator $\hat{S}^2$ in the form

$$\hat{S}^2 = -\frac{N(N-4)}{4} + \sum_{i<j} \hat{P}_{ij}^{\sigma} \quad , \tag{7.29}$$

where $N = N_\alpha + N_\beta$ is the total number of electrons (N_α and N_β being those of electrons with spins α and β, respectively) and operator $\hat{P}_{ij}^{\sigma}$ interchanges the spin coordinates of electrons i and j.[†] A simple derivation shows that the expectation value of this operator for a *single determinant* wave function is given by [91]

$$\langle \hat{S}^2 \rangle = \frac{1}{2}N + \frac{1}{4}(N_\alpha - N_\beta)^2 - \sum_\mu (\mathbf{P}^\alpha \mathbf{S} \mathbf{P}^\beta \mathbf{S})_{\mu\mu} \quad . \tag{7.30}$$

(This formula is routinely used in all standard quantum chemical programs from Gaussian-70 until date.)

One has the identities $N_\alpha = \sum_\mu (\mathbf{P}^\alpha \mathbf{S})_{\mu\mu}$, $N_\beta = \sum_\mu (\mathbf{P}^\beta \mathbf{S})_{\mu\mu}$, thus (7.30) may be written as

$$\langle \hat{S}^2 \rangle = \frac{1}{2}\left[\sum_\mu (\mathbf{P}^\alpha \mathbf{S})_{\mu\mu} + \sum_\mu (\mathbf{P}^\beta \mathbf{S})_{\mu\mu}\right] + \frac{1}{4}\left[\sum_\mu (\mathbf{P}^\alpha \mathbf{S})_{\mu\mu} - \sum_\mu (\mathbf{P}^\beta \mathbf{S})_{\mu\mu}\right]^2 - \sum_\mu (\mathbf{P}^\alpha \mathbf{S} \mathbf{P}^\beta \mathbf{S})_{\mu\mu} \quad . \tag{7.31}$$

By using the spin density matrix $\mathbf{P}^s = \mathbf{P}^\alpha - \mathbf{P}^\beta$, utilizing the idempotency property $(\mathbf{P}^\alpha \mathbf{S})^2 = \mathbf{P}^\alpha \mathbf{S}$, $(\mathbf{P}^\beta \mathbf{S})^2 = \mathbf{P}^\beta \mathbf{S}$ of the density matrices and the identity $Tr(\mathbf{AB}) = Tr(\mathbf{BA})$, we get

$$\langle \hat{S}^2 \rangle = \frac{1}{2}\left[\sum_\mu (\mathbf{P}^\alpha \mathbf{S} \mathbf{P}^\alpha \mathbf{S})_{\mu\mu} + \sum_\mu (\mathbf{P}^\beta \mathbf{S} \mathbf{P}^\beta \mathbf{S})_{\mu\mu}\right] + \frac{1}{4}\left[\sum_\mu (\mathbf{P}^s \mathbf{S})_{\mu\mu}\right]^2 - \frac{1}{2}\left[\sum_\mu (\mathbf{P}^\alpha \mathbf{S} \mathbf{P}^\beta \mathbf{S})_{\mu\mu} + \sum_\mu (\mathbf{P}^\beta \mathbf{S} \mathbf{P}^\alpha \mathbf{S})_{\mu\mu}\right] \quad . \tag{7.32}$$

---

*Parts of this section are reprinted from my papers [88], [89] with permission from Elsevier.

[†]One can obtain this result by investigating how operator $\hat{\vec{s}}_i \hat{\vec{s}}_j$ acts on the different products of spin-functions $\alpha(i)\alpha(j)$, $\beta(i)\alpha(j)$, $\alpha(i)\beta(j)$, and $\beta(i)\beta(j)$. (It follows from the definition of the Pauli matrices that $(\hat{\vec{s}}_i)^2 = \frac{3}{4}\hat{1}_i$, $\hat{1}_i$ being the unity operator with respect to the i-th electron.)

The first two and last two terms can be combined to give $\mathbf{P}^s\mathbf{S}\mathbf{P}^s\mathbf{S}$:

$$\langle \hat{S}^2 \rangle = \frac{1}{2}\sum_{\mu}(\mathbf{P}^s\mathbf{S}\mathbf{P}^s\mathbf{S})_{\mu\mu} + \frac{1}{4}\left[\sum_{\mu}(\mathbf{P}^s\mathbf{S})_{\mu\mu}\right]^2 . \tag{7.33}$$

This result indicates that *there are no spins in systems described by a closed shell RHF wave function* for which $\mathbf{P}^s \equiv \mathbf{0}$, so requirement (i) above is satisfied.

Comparing Eq. (7.33) with the first line of Eq. (7.9), one gets [72]

$$\langle S^2 \rangle = \frac{1}{2}\Delta + \frac{1}{4}\left[\sum_{\mu}(\mathbf{P}^s\mathbf{S})_{\mu\mu}\right]^2 = \frac{1}{2}\Delta + \frac{1}{4}(N_\alpha - N_\beta)^2 , \tag{7.34}$$

establishing a very close relationship between the expectation value $\langle S^2 \rangle$ and the effective number of unpaired electrons Δ in the single determinant case. In particular, for a UHF determinant for which $N_\alpha = N_\beta$ they differ only by a factor of $\frac{1}{2}$.

However, Eqs. (7.33), (7.34) do not satisfy requirement (iii). In fact, the term with the square of the sum is simply a product of two sums:

$$\left[\sum_{\mu}(\mathbf{P}^s\mathbf{S})_{\mu\mu}\right]^2 = \sum_{\mu}(\mathbf{P}^s\mathbf{S})_{\mu\mu}\sum_{\nu}(\mathbf{P}^s\mathbf{S})_{\nu\nu} \tag{7.35}$$

and when one systematizes the terms with subscripts μ and ν according to the centers to which the respective basis orbitals belong, then two-center terms will also appear, in contradiction with the requirement (iii).

According to the discussion on page 106, if there is only one electron in the system, then matrix $2\mathbf{D} - \mathbf{DSD}$ equals the spin density matrix $\mathbf{P}^s$. In that case the spin density equals the total density and also the density of effectively unpaired electrons ρ_u, of course. Thus, taking into account Eq. (7.11), the proper description of the single electron case requires that the formula used for decomposing $\langle \hat{S}^2 \rangle$ should reduce to

$$\langle \hat{S}^2 \rangle \Rightarrow \frac{3}{4}\int \varrho_s(\vec{r})dv = \frac{3}{4}\int \varrho_u(\vec{r})dv = \frac{3}{4}\Delta \tag{7.36}$$

in the case of a *single electron.* In that case it will represent a sum of only one-center terms, without any spurious self-coupling terms. (The same formula remains applicable also for a single unpaired electron outside the closed shells.)

To achieve that goal, we add and extract $\frac{1}{4}\Delta$ at the right-hand side of Eq. (7.34), and use Eq. (7.5), utilizing the idempotency properties of $\mathbf{P}^{\alpha}$ and $\mathbf{P}^{\beta}$:

$$\begin{aligned}\langle \hat{S}^2 \rangle &= \frac{3}{4}\Delta + \frac{1}{4}[Tr(\mathbf{P}^s\mathbf{S})]^2 - \frac{1}{4}\Delta \\ &= \frac{3}{4}\Delta + \frac{1}{4}[Tr(\mathbf{P}^s\mathbf{S})]^2 - \frac{1}{4}Tr\left[(\mathbf{P}^s\mathbf{S})^2\right]\end{aligned} \tag{7.37}$$

Explicit expansion in terms of the orbital coefficients shows that the last two terms cancel *term by term* if there is only a single unpaired electron. In fact, we have

$$\begin{aligned}[Tr(\mathbf{P}^s\mathbf{S})]^2 &= \sum_{A,B}\sum_{\mu\in A}\sum_{\nu\in B}(\mathbf{P}^s\mathbf{S})_{\mu\mu}(\mathbf{P}^s\mathbf{S})_{\nu\nu} \\ &= \sum_{A,B}\sum_{\mu\in A}\sum_{\rho} c_\mu c_\rho^* S_{\rho\mu} \sum_{\nu\in B}\sum_{\tau} c_\nu c_\tau^* S_{\tau\nu}\end{aligned} \tag{7.38}$$

and

$$\begin{aligned}Tr\left[(\mathbf{P}^s\mathbf{S})^2\right] &= \sum_{A,B}\sum_{\mu\in A}\sum_{\nu\in B}(\mathbf{P}^s\mathbf{S})_{\mu\nu}(\mathbf{P}^s\mathbf{S})_{\nu\mu} \\ &= \sum_{A,B}\sum_{\mu\in A}\sum_{\nu\in B}\sum_{\rho,\tau} c_\mu c_\rho^* S_{\rho\nu} c_\nu c_\tau^* S_{\tau\mu} \\ &= \sum_{A,B}\sum_{\mu\in A}\sum_{\tau} c_\mu c_\tau^* S_{\tau\mu} \sum_{\nu\in B}\sum_{\rho} c_\nu c_\rho^* S_{\rho\nu}\end{aligned} \tag{7.39}$$

Every term μ, ν is equal in these two sums, as ρ and τ are interchangeable summation indices.

By grouping the terms of Eq. (7.37) by the atoms and pairs of atoms, one obtains the decomposition of the expectation value as a sum of atomic and diatomic components [89]:

$$\langle \hat{S}^2 \rangle = \sum_A \langle \hat{S}^2 \rangle_A + \sum_{\substack{A,B \\ A\neq B}} \langle \hat{S}^2 \rangle_{AB} \tag{7.40}$$

with

$$\langle \hat{S}^2 \rangle_A = \frac{3}{4}\sum_{\mu\in A}[2\mathbf{DS} - (\mathbf{DS})^2)]_{\mu\mu} + \frac{1}{4}(p_A^s)^2 - \frac{1}{4}\sum_{\mu,\nu\in A}(\mathbf{P}^s\mathbf{S})_{\mu\nu}(\mathbf{P}^s\mathbf{S})_{\nu\mu} \tag{7.41}$$

and

$$\langle \hat{S}^2 \rangle_{AB} = \frac{1}{4}p_A^s p_B^s - \frac{1}{4}\sum_{\mu\in A}\sum_{\nu\in B}(\mathbf{P}^s\mathbf{S})_{\mu\nu}(\mathbf{P}^s\mathbf{S})_{\nu\mu} \tag{7.42}$$

where p_A^s is the gross spin-population of the atom, defined in Eq. (7.7).

The condition (iv) defined on page 115 is not relevant for a UHF wave function which does not correspond to a pure spin state (pure $\hat{S}^2$ eigenfunction) but its fulfillment is to be requested for high-spin ROHF determinant wave functions, for which the lower spin component(s) do not represent a single determinant and should be treated, therefore, alongside the correlated wave functions. Then the $\langle \hat{S}^2 \rangle$ components determined for the high-spin state as a single determinant and for the low spin-state, that is not a single determinant, should be the same, of course.

Calculations of local spins (one- and two-center contributions to $\langle \hat{S}^2 \rangle$) for single determinant wave functions can be performed by using our free program BO-SPIN-2 [92]. It is to be noted here that in the formulae of our publications earlier than 2012 (as well as in our earlier program BO-SPIN) the requirement (iii) pertinent to the case of a single (single unpaired) electron had not yet been accounted for, so they are obsolete and give results different from those discussed here.

The data in Table 7.1 do completely agree with the expectations. The H_2^+ ion is a one-electron system exhibiting the behavior required when condition (iii) was introduced. Systems Li_2^+ and NO practically also behave as systems in which there is a single electron outside the closed shells; in accord with that the diatomic $\langle \hat{S}^2 \rangle$ components are negligible. The significant spin-polarization shown by the allyl radical at the UHF level is also what one could expect. The results obtained for the alternant hydrocarbons for which the UHF method permits to take into account some electron correlation, as it gives some energy lowering as compared with the RHF one, are also interesting. The "splitting" of the orbitals occupied by spins α and β is accompanied by the appearance of some non-negligible local spins. It is not surprising that the largest local spins are observed for cyclobutadiene: it is known [93,94] that the exact π-electron ground state of the hypothetical square cyclobutadiene is a "molecular antiferromagnet" with four almost fully localized spins on the four corners; that is qualitatively well reproduced at the UHF level. (For the true geometry of the cyclobutadiene with alternating single and double bonds the UHF method clearly overestimates this antiferromagnetic character of the wave function—but this is the price paid for some energy lowering as compared with the RHF method.)

Figure 7.1 shows the total spin-square and the atomic spin-squares of the LiH molecule calculated at the UHF level of theory. Around the equilibrium interatomic distance there is no UHF solution differing from the RHF, so all spins are zero. The genuine UHF solution appears in the bifurcation point

Table 7.1
$\langle \hat{S}^2 \rangle$ components of different molecules calculated by the UHF method (cc-pVTZ basis set)

Molecule	Component of $\langle \hat{S}^2 \rangle$			
H_2^+ ion	$\langle \hat{S}^2 \rangle_1$	$\langle \hat{S}^2 \rangle_{12}$		
(R-independent)	0.375	0.0		
Li_2^+ ion	$\langle \hat{S}^2 \rangle_1$	$\langle \hat{S}^2 \rangle_{12}$		
$R_e = 3.1336$Å	0.3750	0.0000		
$R = 1.5$Å	0.3703	0.0047		
H_2 ($R = 1.5$ Å)	$\langle \hat{S}^2 \rangle_1$	$\langle \hat{S}^2 \rangle_{12}$		
	0.4326	-0.1438		
NO	$\langle \hat{S}^2 \rangle_N$	$\langle \hat{S}^2 \rangle_O$	$\langle \hat{S}^2 \rangle_{NO}$	
	0.5730	0.1961	-0.0016	
allyl radical	$\langle \hat{S}^2 \rangle_1$	$\langle \hat{S}^2 \rangle_2$	$\langle \hat{S}^2 \rangle_{12}$	$\langle \hat{S}^2 \rangle_{13}$
	0.5170	0.2089	-0.1510	0.2049
trans-butadiene	$\langle \hat{S}^2 \rangle_1$	$\langle \hat{S}^2 \rangle_2$	$\langle \hat{S}^2 \rangle_{12}$	$\langle \hat{S}^2 \rangle_{23}$
	0.2258	0.1595	-0.0844	-0.0763
	$\langle \hat{S}^2 \rangle_{13}$	$\langle \hat{S}^2 \rangle_{14}$		
	0.0866	-0.1028		
cyclobutadiene D_{2h}*	$\langle \hat{S}^2 \rangle_1$	$\langle \hat{S}^2 \rangle_{12}$	$\langle \hat{S}^2 \rangle_{14}$	$\langle \hat{S}^2 \rangle_{13}$
	0.4107	-0.1873	-0.1922	0.1863
cyclobutadiene D_{4h}	$\langle \hat{S}^2 \rangle_1$	$\langle \hat{S}^2 \rangle_{12}$	$\langle \hat{S}^2 \rangle_{13}$	
	0.5716	-0.2596	0.2536	
benzene	$\langle \hat{S}^2 \rangle_1$	$\langle \hat{S}^2 \rangle_{12}$	$\langle \hat{S}^2 \rangle_{13}$	$\langle \hat{S}^2 \rangle_{14}$
	0.1223	-0.0575	0.0588	-0.0590

*At the RHF minimum. (The UHF minimum corresponds to the D_{4h} symmetry.)
†[89] with permission from Elsevier.

at ~2.214 Å and then the overall $\langle \hat{S}^2 \rangle$ and both atomic spin-squares quickly approach the limiting values of 1 and 0.75, respectively, corresponding to the dissociation into two doublet atoms forming a 1:1 mixture of singlet and triplet. (The actual values are about 0.98 and 0.73–0.74 at the distance of

4 Å.) It may be noted that the curves corresponding to the spins on the Li and H atoms are close but not identical, showing that the problem can nearly but not strictly be reduced to a two-electrons—two-orbitals one. (In a two-electrons—two-orbitals problem the two atomic spins would be identical as a consequence of the symmetry of Wiberg's b_μ parameter Eq. (6.54) with respect to the value $q_\mu = 1$.)

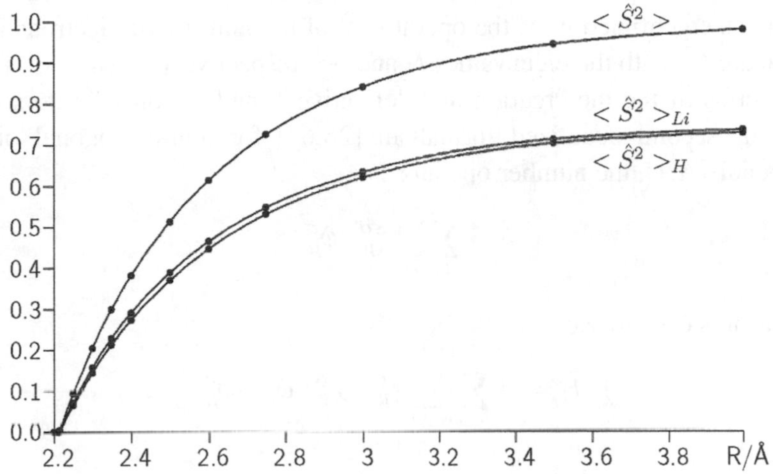

Figure 7.1
Distance dependence of the total spin-square and of atomic spin-squares of the LiH molecules obtained at the UHF level of theory by using 6-31G** basis set.

In the single determinant case requirement (ii) is relevant only if one combines two open-shell determinant wave functions into a resulting overall single determinant. In the asymptotic case the overlap matrix is block-diagonal (the interatomic matrix elements vanish), therefore the intraatomic matrix elements $(\mathbf{P}^s\mathbf{S})_{\mu\nu}$ coincide—may be up to a sign—with those one has for the free atom, which provides the fulfillment of requirement (ii)—as we have seen in the examples of the UHF treatment of molecule H_2 and in the UHF calculations performed for LiH.

7.3.2 Local spins for correlated wave functions

In that case we again start from formula (7.29), explicitly considering every term in it as an operator:

$$\hat{S}^2 = -\frac{1}{4}\hat{N}^2 + \hat{N} + \sum_{i<j} \hat{P}_{ij}^{\sigma} \ . \tag{7.43}$$

This does not represent any limitation because every N-electron wave function is an eigenfunction of the operator $\hat{N}$ of the number of electrons and of its square $\hat{N}^2$ with the eigenvalues N and N^2, respectively.

In terms of the the creation and "effective" annihilation operators of the "mixed" second quantized formalism [26,67] for non-orthogonal orbitals (Appendix A.1), the number operator is

$$\hat{N} = \sum_{\mu}\sum_{\sigma} \hat{\chi}_{\mu}^{\sigma+} \hat{\varphi}_{\mu}^{\sigma-} \ , \tag{7.44}$$

and, as it is easy to see,

$$\sum_{i<j} \hat{P}_{ij}^{\sigma} = \tfrac{1}{2} \sum_{\mu,\nu} \sum_{\sigma,\sigma'} \hat{\chi}_{\mu}^{\sigma'+} \hat{\chi}_{\nu}^{\sigma+} \hat{\varphi}_{\nu}^{\sigma'-} \hat{\varphi}_{\mu}^{\sigma-} \ . \tag{7.45}$$

(Every pair of spins is interchanged and a factor of $\frac{1}{2}$ is introduced to avoid double counting.)

Using these formulae, one readily obtains the second quantized expression of operator $\hat{S}^2$ for the case of a non-orthogonal basis as*

$$\begin{aligned} \hat{S}^2 = \frac{3}{4} \sum_{\mu} \left(\hat{\chi}_{\mu}^{\alpha+} \hat{\varphi}_{\mu}^{\alpha-} + \hat{\chi}_{\mu}^{\beta+} \hat{\varphi}_{\mu}^{\beta-} \right) + \frac{1}{4} \sum_{\mu,\nu} \Big(2\hat{\chi}_{\mu}^{\beta+} \hat{\chi}_{\nu}^{\alpha+} \hat{\varphi}_{\mu}^{\beta-} \hat{\varphi}_{\nu}^{\alpha-} \\ - \hat{\chi}_{\mu}^{\alpha+} \hat{\chi}_{\nu}^{\alpha+} \hat{\varphi}_{\mu}^{\alpha-} \hat{\varphi}_{\nu}^{\alpha-} - \hat{\chi}_{\mu}^{\beta+} \hat{\chi}_{\nu}^{\beta+} \hat{\varphi}_{\mu}^{\beta-} \hat{\varphi}_{\nu}^{\beta-} - 4\hat{\chi}_{\mu}^{\alpha+} \hat{\chi}_{\nu}^{\beta+} \hat{\varphi}_{\mu}^{\beta-} \hat{\varphi}_{\nu}^{\alpha-} \Big) \ . \end{aligned} \tag{7.47}$$

*Alternatively, one can, of course, start from the expressions

$$\hat{S}_x = \tfrac{1}{2} \sum_{\mu} \left(\hat{\chi}_{\mu}^{\beta+} \varphi_{\mu}^{\alpha-} + \hat{\chi}_{\mu}^{\alpha+} \varphi_{\mu}^{\beta-} \right), \quad \hat{S}_y = \frac{i}{2} \sum_{\mu} \left(\hat{\chi}_{\mu}^{\beta+} \varphi_{\mu}^{\alpha-} - \hat{\chi}_{\mu}^{\alpha+} \varphi_{\mu}^{\beta-} \right),$$

$$\hat{S}_z = \tfrac{1}{2} \sum_{\mu} \left(\hat{\chi}_{\mu}^{\alpha+} \varphi_{\mu}^{\alpha-} - \hat{\chi}_{\mu}^{\beta+} \varphi_{\mu}^{\beta-} \right) . \tag{7.46}$$

for the operators of the components of the total spin, that can be obtained directly from the Pauli matrices, and utilizing the anticommutation property $\{\hat{\chi}_{\mu}^{\sigma+} ; \hat{\varphi}_{\nu}^{\sigma'-}\} = \delta_{\sigma\sigma'} \delta_{\mu\nu}$, one arrives at the same Eq. (7.47) for the $\hat{S}^2 = \hat{S}_x^2 + \hat{S}_y^2 + \hat{S}_z^2$.

The terms of (7.47) could be grouped according to the individual atoms and pairs of atoms. However, that seemingly straightforward approach would lead to the decomposition of the expectation value $\langle\hat{S}^2\rangle$ in the manner of Refs. 84–86, which does not satisfy condition (i) on page 115, and thus does not fit our physical picture of covalent systems. For that reason we should adjust our treatment to fulfill conditions (i) to (iv), analogously as done above in the single determinant case.

The expectation value of the string $\hat{\chi}_\mu^{\sigma+}\hat{\varphi}_\nu^{\sigma'-}$ can be expressed [26,67] (c.f. Appendix A.1) through the LCAO "density matrix" $\mathbf{P}^\sigma$ ($\sigma=\alpha,\ \beta$) as:

$$\langle\hat{\chi}_\mu^{\sigma+}\hat{\varphi}_\nu^{\sigma'-}\rangle = \delta_{\sigma\sigma'}(\mathbf{P}^\sigma\mathbf{S})_{\nu\mu} \quad , \tag{7.48}$$

so the expectation value of the first sum in Eq. (7.47) is simply $\frac{3}{4}\sum_\mu(\mathbf{DS})_{\mu\mu}$. The expectation values of the other operator strings lead to different second-order density matrix elements that, in turn, can be written down as their components valid in the single determinant case, expressible through the first-order density matrix elements, and the respective cumulants. But before using that, we shall somewhat rearrange these terms by adding and subtracting some terms and utilizing the anticommutation properties of the operator strings and of the fact that μ and ν are interchangeable summation indices. We write

$$\begin{aligned}
\langle\hat{S}^2\rangle &= \frac{3}{4}\sum_\mu(\mathbf{DS})_{\mu\mu} + \frac{1}{4}\sum_{\mu,\nu}\Big(2\langle\hat{\chi}_\mu^{\beta+}\hat{\chi}_\nu^{\alpha+}\hat{\varphi}_\mu^{\beta-}\hat{\varphi}_\nu^{\alpha-}\rangle \\
&\quad -\langle\hat{\chi}_\mu^{\alpha+}\hat{\chi}_\nu^{\alpha+}\hat{\varphi}_\mu^{\alpha-}\hat{\varphi}_\nu^{\alpha-}\rangle - \langle\hat{\chi}_\mu^{\beta+}\hat{\chi}_\nu^{\beta+}\hat{\varphi}_\mu^{\beta-}\hat{\varphi}_\nu^{\beta-}\rangle - 4\langle\hat{\chi}_\mu^{\alpha+}\hat{\chi}_\nu^{\beta+}\hat{\varphi}_\mu^{\beta-}\hat{\varphi}_\nu^{\alpha-}\rangle\Big) \\
&= \frac{3}{4}\sum_\mu(\mathbf{DS})_{\mu\mu} - \frac{1}{4}\sum_{\mu,\nu}\Big[2\Big(\langle\hat{\chi}_\mu^{\alpha+}\hat{\chi}_\nu^{\alpha+}\hat{\varphi}_\mu^{\alpha-}\hat{\varphi}_\nu^{\alpha-}\rangle + \langle\hat{\chi}_\mu^{\beta+}\hat{\chi}_\nu^{\beta+}\hat{\varphi}_\mu^{\beta-}\hat{\varphi}_\nu^{\beta-}\rangle \\
&\quad + \langle\hat{\chi}_\mu^{\beta+}\hat{\chi}_\nu^{\alpha+}\hat{\varphi}_\mu^{\alpha-}\hat{\varphi}_\nu^{\beta-}\rangle + \langle\hat{\chi}_\mu^{\alpha+}\hat{\chi}_\nu^{\beta+}\hat{\varphi}_\mu^{\beta-}\hat{\varphi}_\nu^{\alpha-}\rangle\Big) + \langle\hat{\chi}_\mu^{\alpha+}\hat{\chi}_\nu^{\alpha+}\hat{\varphi}_\nu^{\alpha-}\hat{\varphi}_\mu^{\alpha-}\rangle \\
&\quad + \langle\hat{\chi}_\mu^{\beta+}\hat{\chi}_\nu^{\beta+}\hat{\varphi}_\nu^{\beta-}\hat{\varphi}_\mu^{\beta-}\rangle + \langle\hat{\chi}_\mu^{\beta+}\hat{\chi}_\nu^{\alpha+}\hat{\varphi}_\nu^{\alpha-}\hat{\varphi}_\mu^{\beta-}\rangle + \langle\hat{\chi}_\mu^{\alpha+}\hat{\chi}_\nu^{\beta+}\hat{\varphi}_\nu^{\beta-}\hat{\varphi}_\mu^{\alpha-}\rangle\Big]
\end{aligned} \tag{7.49}$$

Now we separate out those parts of the expectation values of the operator strings (i.e., second-order density matrix elements) that can be expressed through the first density matrix elements and the respective "spin-less" cumulants. We use the relationship

$$\begin{aligned}
\langle\hat{\chi}_\mu^{\sigma+}\hat{\chi}_\nu^{\sigma'+}\hat{\varphi}_\rho^{\sigma''-}\hat{\varphi}_\tau^{\sigma'''-}\rangle &= \delta_{\sigma\sigma'''}\delta_{\sigma'\sigma''}(\mathbf{P}^\sigma\mathbf{S})_{\tau\mu}(\mathbf{P}^{\sigma'}\mathbf{S})_{\rho\nu} \\
&\quad -\delta_{\sigma'\sigma'''}\delta_{\sigma\sigma''}(\mathbf{P}^{\sigma'}\mathbf{S})_{\tau\nu}(\mathbf{P}^\sigma\mathbf{S})_{\rho\mu} + \sum_{\lambda,\eta}\Lambda_{\tau^\sigma\rho^{\sigma'}\lambda^{\sigma''}\eta^{\sigma'''}}S_{\lambda\mu}S_{\eta\nu} \quad ,
\end{aligned} \tag{7.50}$$

and get, taking into account the definition (7.17) of the "spin-less" cumulant as sum of four spin-dependent ones:

$$\begin{aligned}\langle \hat{S}^2 \rangle = & \frac{3}{4}\sum_{\mu}(\mathbf{DS})_{\mu\mu} \\ & +\frac{1}{4}\sum_{\mu\nu}\Big[(\mathbf{P}^\alpha\mathbf{S})_{\mu\mu}(\mathbf{P}^\alpha\mathbf{S})_{\nu\nu}+(\mathbf{P}^\beta\mathbf{S})_{\mu\mu}(\mathbf{P}^\beta\mathbf{S})_{\nu\nu}-2(\mathbf{P}^\alpha\mathbf{S})_{\mu\mu}(\mathbf{P}^\beta\mathbf{S})_{\nu\nu} \\ & -(\mathbf{P}^\alpha\mathbf{S})_{\mu\nu}(\mathbf{P}^\alpha\mathbf{S})_{\nu\mu}-(\mathbf{P}^\beta\mathbf{S})_{\mu\nu}(\mathbf{P}^\beta\mathbf{S})_{\nu\mu}-4(\mathbf{P}^\alpha\mathbf{S})_{\mu\nu}(\mathbf{P}^\beta\mathbf{S})_{\nu\mu}\Big] \\ & -\frac{1}{4}\sum_{\mu,\nu,\rho,\tau}\left[2\Lambda_{\mu\nu\rho\tau}S_{\rho\nu}S_{\tau\mu}+\Lambda_{\mu\nu\rho\tau}S_{\rho\mu}S_{\tau\nu}\right]\end{aligned} \tag{7.51}$$

As discussed above, in order to satisfy condition (iii) pertinent to the case when there is a single (unpaired) electron in the system, one has to separate out the term $\frac{3}{4}\Delta$ in the formula for $\langle \hat{S}^2 \rangle$; then all the other terms should cancel or vanish in the single electron case. To achieve that goal, we add and extract a term $-\frac{3}{4}\sum_{\mu,\nu,\rho,\tau}\Lambda_{\mu\nu\rho\tau}S_{\rho\mu}S_{\tau\nu}$ in the above formula, and rewrite it according to Eq. (7.24) as

$$-\frac{3}{4}\sum_{\mu,\nu,\rho,\tau}\Lambda_{\mu\nu\rho\tau}S_{\rho\mu}S_{\tau\nu}=\frac{3}{8}\Delta-\frac{3}{8}\sum_{\mu,\nu}(\mathbf{P}^s\mathbf{S})_{\mu\nu}(\mathbf{P}^s\mathbf{S})_{\nu\mu} \tag{7.52}$$

Performing simple manipulations in which the differences $\mathbf{P}^\alpha-\mathbf{P}^\beta$ are combined into the spin-density matrix $\mathbf{P}^s$, we get finally

$$\begin{aligned}\langle \hat{S}^2 \rangle = & \frac{3}{4}\Delta+\frac{1}{4}\sum_{\mu,\nu}(\mathbf{P}^s\mathbf{S})_{\mu\mu}(\mathbf{P}^s\mathbf{S})_{\nu\nu}-\frac{1}{4}\sum_{\mu,\nu}(\mathbf{P}^s\mathbf{S})_{\mu\nu}(\mathbf{P}^s\mathbf{S})_{\nu\mu} \\ & +\frac{1}{2}\sum_{\mu,\nu,\rho,\tau}\left[\Lambda_{\mu\nu\rho\tau}-\Lambda_{\mu\nu\tau\rho}\right]S_{\rho\mu}S_{\tau\nu}\end{aligned} \tag{7.53}$$

The first three terms of this equation are identical to those of the last line of Eq. (7.37) giving the expectation value $\langle \hat{S}^2 \rangle$ for a single determinant wave function, while the remaining ones contain the cumulants that vanish in the single determinant case. Grouping the terms according to the individual atoms and pairs of atoms we obtain the atomic and diatomic components we were looking for:

$$\begin{aligned}\langle \hat{S}^2 \rangle_A = & \frac{3}{4}\sum_{\mu\in A}[2\mathbf{DS}-(\mathbf{DS})^2)]_{\mu\mu}+\frac{1}{4}(p_A^s)^2-\frac{1}{4}\sum_{\mu,\nu\in A}(\mathbf{P}^s\mathbf{S})_{\mu\nu}(\mathbf{P}^s\mathbf{S})_{\nu\mu} \\ & +\frac{1}{2}\sum_{\mu,\nu\in A}\sum_{\tau,\rho}\left[\Lambda_{\mu\nu\rho\tau}-\Lambda_{\mu\nu\tau\rho}\right]S_{\rho\mu}S_{\tau\nu}\end{aligned} \tag{7.54}$$

and

$$\langle \hat{S}^2 \rangle_{AB} = \frac{1}{4} p_A^s p_B^s - \frac{1}{4} \sum_{\mu \in A} \sum_{\nu \in B} (\mathbf{P}^s\mathbf{S})_{\mu\nu} (\mathbf{P}^s\mathbf{S})_{\nu\mu} \tag{7.55}$$
$$+ \frac{1}{2} \sum_{\mu \in A} \sum_{\nu \in B} \sum_{\tau,\rho} \left[\Lambda_{\mu\nu\rho\tau} - \Lambda_{\mu\nu\tau\rho} \right] S_{\rho\mu} S_{\tau\nu}$$

where p_A^s is again the gross spin-population of the atom, defined in Eq. (7.7). The same formulae were derived in our paper [95] in a somewhat different manner.

Table 7.2
CISD atomic $\langle \hat{S}^2 \rangle_A$ and diatomic $\langle \hat{S}^2 \rangle_{AB}$ values for a set of singlet molecules at optimized geometries (6-31G** basis set)

Molecule		$\langle \hat{S}^2 \rangle_A / \langle \hat{S}^2 \rangle_{AB}$	Molecule		$\langle \hat{S}^2 \rangle_A / \langle \hat{S}^2 \rangle_{AB}$
H_2	H	0.036	C_2H_6	C	0.199
	H-H	-0.036		H	0.024
Li_2	Li	0.156		C-C	-0.122
	Li-Li	-0.156		C-H	-0.069
Be_2	Be	0.175		C..H	0.034
	Be-Be	-0.175		H-H	0.018
HF	H	0.006		H..H	-0.015
	F	0.006	C_2H_4	C	0.056
	H-F	-0.006		H	0.024
H_2O	H	0.013		C-C	-0.094
	O	0.013		C-H	-0.036
	O-H	-0.007		C..H	0.055
	H..H	-0.006		H-H	-0.002
NH_3	N	0.061		$H..H_{cis}$	-0.026
	H	0.019		$H..H_{trans}$	-0.014
	N-H	-0.020	C_2H_2	C	-0.139
	H..H	0.000		H	0.019
CH_4	C	0.320		C-C	0.147
	H	0.026		C-H	0.048
	C-H	-0.080		C..H	-0.056
	H..H	0.018		H..H	-0.006

*[95] permission from Royal Chemical Society.

Tables 7.2 and 7.3 contain some numerical values calculated for CISD and CAS-SCF wave functions, respectively, by using 6-31G** basis set [95]. Most numbers correspond to the expectations, except the relatively large local spin on the carbon atom and relatively small one on the hydrogens in the methane molecule, and the quite unexpected negative carbon $\langle \hat{S}^2 \rangle$ contributions in the acetylene. As to the latter, the detailed study performed in Ref. 95 indicates that it can be attributed to a defect of the Mulliken-type analysis we performed, as no negative atomic $\langle \hat{S}^2 \rangle$ components emerge if one performs the analysis in a Löwdin-orthogonalized basis or in the "fuzzy atoms" 3D framework discussed in Section 9.7.

Table 7.3

Atomic $\langle \hat{S}^2 \rangle_C$ and diatomic $\langle \hat{S}^2 \rangle_{CC}$ components at the CAS-SCF/6-31G** level of theory. Active spaces used are (6,6) for benzene, (7,7) for phenyl radical, and (3,3) for the allyl radical

	C_1	C_2	C_3	C_4	C_5	C_6
Benzene						
C_1	0.114	-0.101	0.079	-0.069		
Phenyl						
C_1	0.931	-0.127	0.059	-0.026		
C_2		0.098	-0.055	0.013	-0.029	0.075
C_3			0.061	-0.018	0.001	
C_4				0.024		
Allyl						
C_1	0.145	-0.145				
C_2		0.440	0.151			

*[95] permission from Royal Chemical Society.

8

Energy components in the Hilbert space

Until now we have discussed different effects of chemical significance on the basis of considering quantities related to the different type of densities. Now we shall turn to *energetic* effects—they are of utmost importance for chemistry: the energy of the system is crucial for all phenomena—from molecular geometry to the dynamics of the reactions. However, the total energy is only a *single number*; one has to understand the factors governing its changes under different circumstances. We know that there are quite similar bonds and groups in different molecules, indicating that there are similar energetic effects in them. Our task is to identify and utilize these local parameters: the understanding and interpretation of the results obtained in a quantum chemical calculation can be much facilitated by presenting the total energy as a sum of chemically meaningful components.

One can speak about the existence of a chemical bond between the two atoms of a diatomics if the energy of the system with the nuclei at some finite distance is lower than the sum of the free atomic energies. This definition can intuitively be generalized to multiatomic molecules: one has to invest energy to increase the distance between the chemically bonded atoms as compared with the equilibrium molecular structure. (More precisely, one should speak about the distance of the respective nuclei, of course, as we are working in the framework of the Born–Oppenheimer approximation.) To a great extent, chemists consider the energy of a molecule to be the sum of the atomic energies, energies of the chemical bonds and of the non-bonded interactions. There are also, of course, different collective effects involving more than two atoms (e.g., aromaticity); they, however, can also be treated—at least formally—as combinations (or interferences) of one- and diatomic contributions. Accordingly, we are going to consider different schemes in which the total energy is presented as a sum of terms corresponding to the individual atoms and to the pairs of atoms.

The atomic energies in the molecules are different from those of the free atoms. This is due to different reasons. One of the effects is called "promotion," and is connected with the change of the electronic structure of an atom as it forms chemical bond(s). In some cases this can be rather significant, in particular when the atom enters the molecule in a different electronic configuration than that of the atomic ground state: it is enough to recall here the different hybridization states (sp, sp^2, sp^3) of the carbon atoms in organic molecules—without the change from the ground state electronic configuration s^2p^2 to the promoted sp^3 carbon would be divalent and not four-valent. Another important factor is that the atoms in the molecule are not strictly neutral, but acquire some ionic character: the accumulation or loss of electrons is also connected with changes in the energy. One may say that electrons flow between the atoms to reduce the resulting total molecular energy. Of course, the two-center energy contributions also reflect interactions of different types, like electrostatic, exchange, overlap, etc.

As discussed in Section 3.1, the presence of the three- and four-center integrals makes the energy decomposition in the Hilbert-space analysis a complex task. Before entering that field, we shall briefly discuss an approach that permits to circumvent the problem of multicenter integrals on the basis of *virial theorem.*

8.1 Energy partitioning based on the virial theorem

Unfortunately, for molecular systems treated in the framework of the Born–Oppenheimer approximation, the virial theorem holds in its familiar form only at the equilibrium configuration of the molecules [5] and only either for the exact wave function or for variational ones treated in the limit of complete basis sets.* (That means the Hartree–Fock limit if single determinants are used.) However, while the first condition (equilibrium configuration) cannot be easily avoided, the second requirement (complete basis set) can practically be disregarded, and one can get approximate results of acceptable accuracy.

The general form of the virial theorem in the framework of the Born–Oppenheimer approximation is [5]:

$$2\langle \hat{T}_e \rangle + \langle V \rangle + \sum_{\alpha} \vec{R}_{\alpha} \frac{\partial E}{\partial \vec{R}_{\alpha}} = 0 \, . \tag{8.1}$$

*In Ref. [5] only the case of exact wave function is discussed; the generalization for variational methods has been added in its Russian translation [96]. A generalization of the virial theorem for general (non-variational and/or finite basis) wave functions—leading to the appearance of several additional terms like the sum in Eq. (8.1)—has been considered in the papers [97,98]. We shall not discuss that possibility here, however.

where T_e is the *kinetic energy of the electrons*, V is the *total potential energy* of the system (including the nuclear-nuclear repulsion), E is the total Born–Oppenheimer electronic energy

$$E = \langle \hat{T}_e \rangle + \langle V \rangle \,, \tag{8.2}$$

i.e., again including the nuclear-nuclear repulsion, and $\vec{R}_\alpha$ is the radius-vector of the α-th nucleus. At the equilibrium configuration $\partial E / \partial \vec{R}_\alpha = 0$ and we are left with the familiar

$$2\langle \hat{T}_e \rangle + \langle V \rangle = 0 \,. \tag{8.3}$$

Equations (8.2) and (8.3) can be combined to give

$$E = -\langle \hat{T}_e \rangle \,. \tag{8.4}$$

In practice—in particular if finite basis sets are used—this equality is not fulfilled exactly even at the equilibrium geometry, but often it may be considered a good approximation. Thus, we can write, using the explicit expression of $\langle \hat{T}_e \rangle$ in the case of a finite basis expansion:

$$E \cong -\langle \hat{T} \rangle = \sum_{\mu,\nu} D_{\mu\nu} T_{\nu\mu} = -\sum_{\mu,\nu} D_{\mu\nu} \langle \chi_\nu | -\tfrac{1}{2}\Delta | \chi_\mu \rangle \,. \tag{8.5}$$

Here $D_{\mu\nu}$ is the element of the usual spinless density matrix.

Now we may apply two "atomic resolutions of identity" and write, according to Eq. (3.6):

$$E \cong -\sum_{A,B} \sum_{\mu,\nu} D_{\mu\nu} \langle \chi_\nu | \hat{\rho}_A^\dagger \left(-\tfrac{1}{2}\Delta\right) \hat{\rho}_B | \chi_\mu \rangle \,. \tag{8.6}$$

Introducing the definition Eq. (3.8) of the atomic operators for the Hilbert space analysis—which means systematizing the terms of Eq. (8.5) according to the atoms on which the different AOs are centered—and combining integrals which are equal assuming that real basis orbitals and orbital coefficients are used), we get an approximate energy decomposition in the form [99,100]

$$E \cong -\sum_{A} \sum_{\mu,\nu \in A} D_{\mu\nu} T_{\nu\mu} - 2 \sum_{A<B} \sum_{\mu \in A} \sum_{\nu \in B} D_{\mu\nu} T_{\nu\mu} \,, \tag{8.7}$$

where $T_{\nu\mu} = \langle \chi_\nu | -\frac{1}{2}\Delta | \chi_\mu \rangle$ are the standard kinetic energy integrals. Thus we could present the total energy of the molecule in the form in which it is (approximately) equal to a sum of one- and two-center (mono- and diatomic) contributions, which can be computed pretty easily. Calculations of such type may be performed by our freely available program [101].

As the virial theorem is related to the global quantities, it is hard to give any detailed physical interpretation to the different energy components. However, experience shows that this decomposition reflects extremely well the standard picture of molecules consisting of atoms held together with chemical bonds. Independently of the basis set used, the significant energy components correspond to the individual atoms and to the *chemically bonded* pairs of atoms. As the results in Table 8.1 show, the "virial ratio" $-\langle V \rangle / \langle \hat{T} \rangle$ does not deviate too much from the ideal value of 2 predicted in Eq. (8.3) even for the simplest basis set STO-3G. The quantities corresponding to the individual atoms are significantly higher than the free atomic energies,* the diatomic terms corresponding to the chemically bonded atoms are significant negative numbers, while the non-bonded interactions are rather marginal. Thus the method gives relatively large promotion energies and large (in absolute values) binding energies. A special advantage of this scheme (in addition to its simplicity) is that apparently it permits to unambiguously identify the existence or absence of a true chemical bond: remarkable in this context is the significant B–B diatomic energy component in the diborane molecule, which is in complete agreement with our previous discussion about the three-center bonds.

The intimate connection between the virial-type energy decomposition and the conventional chemical bonds can be understood if one considers the single determinant wave function in terms of localized orbitals φ_i^l, that can be sufficiently well assigned to individual atoms or pairs of atoms. (In practice, the localized orbitals show a significant delocalization only in conjugated, or similar, systems, but even if that happens, it does not destroy our argument.) In fact, the kinetic energy is a *one-electron* quantity, therefore its expectation value can be written in the RHF case as a sum of contributions from the individual MOs:

$$\langle \hat{T} \rangle = 2 \sum_{i}^{occ.} \langle \varphi_i^l | \hat{T} | \varphi_i^l \rangle , \qquad (8.8)$$

and a similar formula holds in the UHF case, too. Now, if the orbitals are indeed localized, then each term $\langle \varphi_i^l | \hat{T} | \varphi_i^l \rangle$ will give significant contributions either to a single atom or to that pair of atoms to the bond of which it actually corresponds.

A significant disadvantage of this approach consists in the fact that Eq. (8.3) is applicable only at the equilibrium configuration of molecular systems, so it cannot be used to discuss *changes* of energy components. Eq. (8.1) has

*Exceptions are the hydrogens in the STO-3G basis set, which uses the STO exponent adjusted to the molecular situation, and therefore the energy of the free hydrogen atom is very poor in this basis.

Table 8.1
Virial energy partitioning results for the ethane and diborane molecules calculated by using STO-3G, 6-31G** and cc-pVTZ basis sets at the equilibrium bond distances

Molecule	Atom, bond	$\Delta E_A^\dagger$, E_{AB} (kcal/mol)		
		STO-3G	6-31G**	cc-pVTZ
Ethane	C	674.9	436.8	527.1
	H	-4.0	45.3	43.3
	C–C	-233.1	-239.2	-277.4
	C–H	-232.2	-245.9	-265.3
	H–H geminal	-0.73	2.08	1.04
	H–H vicinal (2×)	-0.60	-0.64	-0.37
	(1×)	0.90	1.52	1.08
	C–H vicinal	-1.45	3.21	2.30
	Virial ratio	2.00783	2.00049	2.00046
Diborane	B	291.0	303,4	295.4
	$H_{br.}$	-36.1	6.4	51.3
	$H_{term.}$	-22.4	35.0	63.3
	B–B	-69.8	-80.7	-111.6
	B–$H_{br.}$	-83.6	-96.7	-110.5
	B–$H_{term.}$	-203.1	-206.1	-217.0
	$H_{br.}$–$H_{br.}$	-3.0	1.3	-0.50
	$H_{br.}$–$H_{term.}$	-0.81	-0.25	-0.21
	$H_{term.}$–$H_{term.}$	-0.96	-0.51	0.04
	$H_{term.}$–$H'_{term.}$	-0.30	0.39	-0.05
	$H_{term.}$–$H''_{term.}$	0.31	0.72	0.52
	Virial ratio	1.99563	1.99880	2.00044

†Change with respect to the free atom energy.

more general applicability but cannot be used for decomposition. In fact, the terms of the sum formally can be considered as corresponding to the individual atoms. However, if one shifts the origin of the coordinate system by a vector $\vec{R}_0$, which corresponds to replacing every $\vec{R}_\alpha$ by $\vec{R}_\alpha - \vec{R}_0$, then the sum does not change,* but the individual components belonging to the different atoms do. Thus the decomposition with these terms included would not be invariant under the change of the laboratory system of coordinates, and is useless, therefore.

*It is assumed that there are only internal forces in the system; then the forces $\vec{F}_\alpha = \partial E/\partial \vec{R}_\alpha$ sum to zero, thus the sum in Eq. (8.1) remains invariant.

8.2 Atomic promotion energies in molecules

As noted in the previous section, the virial theorem reflects some global behavior of the system, and therefore—although it yields rather reasonable atomic and diatomic energy terms at equilibrium conformations—it cannot be used to analyze the individual local factors influencing the formation of molecules. For that purpose we have to deal with the conventional molecular energy formula containing all the one- and two-electron integrals and also we should go step by step, and consider the different effects separately.*

The state of an atom in the molecule represents a big interest in chemistry—it is enough to recall here the distinction of sp, sp^2, and sp^3 carbon atoms in organic chemistry. While these concepts are indeed hard to strictly quantify in the *ab initio* framework, they work rather well (not to say mysteriously well) in practice. It is obvious that these changes in the state of the atoms do have energetic consequences: for instance, we know that all the sp^n hybridized states of carbon can be deduced from a valence state with the "promoted" electron configuration $(1s)^2(2s)^1(2p)^3$, and not from the ground state electron configuration $(1s)^2(2s)^2(2p)^2$ of the free carbon atom.

In Section 5.1 we discussed that for each atom in the molecule one can calculate a set of effective atomic orbitals, the occupation numbers of which measure the extent to which the different effective AOs actually participate in the molecular SCF wave function. We have also seen that these effective AOs in all cases could be grouped as the orbitals forming an "effective minimal basis" and all the others having minor or even negligible importance. It is, therefore, a plausible idea to consider the orbitals of the effective minimal basis as being those by which the atom enters the molecular wave function. Thus a determinant wave function constructed by using these orbitals and assigning them appropriate *integer* occupation numbers (one or two) should be able to represent well the *promoted electronic state* of the atom under consideration in the given molecule. Such an atomic "reference state" reflects not only the change of the electron configuration but also the *distortion* of the orbitals as compared with the free atomic ones.

The occupation numbers to be used to construct these promoted atomic wave functions can be obtained by performing a population analysis of the wave function of the molecule studied. In order to decide whether an effective AO should be considered doubly occupied, singly occupied, or empty

*Part of this section is reprinted from my paper [102] with permission from Elsevier.

in the promoted atomic wave function, one should transform the density and overlap matrices to the basis of the effective AO's and calculate Mulliken's gross atomic populations of the individual effective atomic orbitals. Rounding them to the nearest integer, one gets the occupation number to be used for the given orbital. Adding these occupation numbers, one usually gets as many electrons as is the number of electrons in the neutral atoms; in some cases, however—e.g., for molecules containing hypervalent sulfur atom—ionic states are obtained, giving very important information about the electronic structure of the molecule.*

As different spin couplings within the same atom are of no chemical significance, a high-spin single determinant wave function with all the singly occupied orbitals having the same (α) spin in accord with Hund's rule is constructed. The energy of such a determinant wave function is then identified with the energy of the promoted atom entering the molecule. Accordingly, the promotion energy of the atom is equal to the difference of this promoted energy and that of the atomic ground state ROHF wave function [102].

It is to be stressed that the "reference state" obtained by using integer occupation numbers represents only a hypothetical intermediate entity which is devoted to facilitate the analysis, but does not fully describe the actual state of an atom *within* the molecule. This reference state describes the change in the orbital configuration scheme of the atom (if any) and the orbital deformations occurring in the molecule, but does not reflect the different delocalization and electron transfer effects present in the molecule. Therefore the promotion energy calculated in the manner described cannot be identified with the actual energy of the atom in the molecule. Thus, for instance, the "reference state" does not account also for the fact that $2s$ orbitals of a carbon are lower in energy than the $2p$ ones, and have, in accord with that, larger occupations in the resulting wave functions—a fact, not described if one considers only localized two-center bonds formed by idealized linear, trigonal, or tetrahedral hybrids.† This difference in the energies of the s and p atomic orbitals has an important effect on determining the finer details of molecular structure, as will be discussed in the next section.

*In some exceptional cases it may be necessary to adjust the occupation numbers obtained simply by rounding the Mulliken gross populations to the nearest integer, in order to ensure the overall neutrality of the molecule. For instance, for an electronegative atom it may be necessary to consider an orbital with a Mulliken's gross orbital population of, say, 1.52 as singly occupied (and not doubly occupied) in the reference atomic state.

†Also, the simple high-spin atomic function does not account for the appearance of the ionic configurations in the wave function that are connected with the bond formation, which are of great importance from the point of view of calculating one-center energy components, as will be discussed in Section 8.5.

Table 8.2
Atomic promotion energies calculated for selected molecules by using the 6-31G** and cc-pVTZ basis sets

Molecule	Atom	Promotion energy (kcal/mol) 6-31G**	cc-pVTZ
H_2	H	12.21	12.53
Water	O	151.69	149.34
	H	46.10	36.29
Methane	C	96.25	91.83
	H	14.39	15.86
Ethane	C	96.61	89.59
	H	13.17	14.65
Propane	C_1	95.90	88.25
	C_2	97.34	87.35
	H (CH_3)	13.57	15.88
	(2×)	12.78	14.56
	H (CH_2)	12.35	13.98
Butane	C_1	96.44	88.27
	C_2	96.64	86.33
	H (CH_3)	13.57	16.10
	(2×)	12.85	14.66
	H (CH_2)	12.03	14.01
Isobutane	C_1	95.24	87.22
	C_2	97.67	85.48
	H (CH_3)	12.46	14.57
	(2×)	13.33	15.77
	H (CH)	11.79	13.92
Ethylene	C	102.45	100.45
	H	18.32	21.29
Acetylene	C	116.42	116.14
	H	33.55	30.31
Methanol	C	126.29	119.56
	O	141.10	156.25
	H (CH_3)	16.38	15.87
	(2×)	11.41	11.38
	H (OH)	46.67	32.70
Ammonia	N	171.04	192.79
	H	31.62	25.04

* [102] with permission from Elsevier.

8.3 Role of hybridization and the VSEPR rules

At this point we make a detour from our main topic of *a posteriori* analysis of the calculations to a *qualitative* predictive theory. The fact that the *a posteriori* analysis of calculations always leads to the same number of significantly populated effective atomic orbitals as the number of the functions in a classical minimal basis set, gives a confidence that the theoretical considerations utilizing such basis sets may also be of relevance.

Following the paper [103], we shall discuss two empirical rules of the so-called "valence shell electron pair repulsion" (VSEPR) model [104], as they are applicable to characterizing valence angles formed by the main group elements.* Considering a "central atom" and its "ligands," one usually observes that

1. The lone pairs occupy a larger solid angle than the bonding orbitals; *and*
2. The space requirement of a bonding orbital is reduced if the ligand is more electronegative than the central atom.

In terms of the valence angles θ, the first rule may be formulated as the inequality

$$\theta_{ll'} > \theta_{lb} > \theta_{bb'} \;, \tag{8.9}$$

where l, l' and b, b' denote the different lone pairs and bonding orbitals of the central atom. Obviously, the second rule gives a refinement of the first.

These rules may easily be understood on the basis of the variation principle and of the difference between the energies of the atomic s- and p-orbitals, by using a very simple hybridization model.

Let us consider an approximation in which the wave function of the molecule is built up by using doubly occupied *strictly localized*—i.e., one- or two-center—molecular orbitals. This is, of course, not a very precise description, but the example described in the recent paper [105] also indicates that it may be adequate to account for the most important stereochemical effects. Anyway, we are looking for general tendencies, and they must manifest already at such a simple level of theory—otherwise the presence of different finer effects depending on the variable environment would prevent the existence of such general regularities. (The same holds for our other simplifications and approximations to be introduced.)

*No strained or hypervalent systems will be considered.

We consider a conventional minimal s-p basis on the atoms considered. the two-center bonding orbitals φ_i can be written as

$$\varphi_i = c_i h_i + d_i \kappa_i \ , \qquad (8.10)$$

where h_i is a hybrid atomic orbital on the central atom directed to the ligand considered, κ_i is the respective orbital of the ligand, and c_i, d_i are the MO coefficients. The lone pairs are considered to be completely localized on the central atom; they also may be described by Eq. (8.10), assuming the values $c_i = 1$, $d_i = 0$. (Then, of course, κ_i disappears from the expression.)

The coefficients c_i, d_i are basically determined by the relative electronegativities of the central atom and of the ligand: a more electronegative ligand attracts the electrons which results in a smaller c_i value. In the following we shall consider the coefficients c_i to be constants determined by the nature of the ligands, and shall concentrate on the factors having immediate influence on the valence angles between the hybrids h_i centered on the given atom. (In other words, we shall not take explicitly into account that the change in the hybridization state of the central atom can lead to some changes in the coefficients c_i, d_i.)

The hybrid orbital h_i is defined as

$$h_i = a_i s + b_i p_i \qquad (a_i^2 + b_i^2 = 1) \ , \qquad (8.11)$$

where s and p_i are valence s and p orbitals of the central atom, and the orbital p_i is oriented toward the respective bond or lone pair direction. (If the orbital p_i is directed to the ligand by its positive half, then $a_i, b_i \geq 0$.) The hybrids h_i and h_j are assumed orthonormalized, for which one has to require [106]:

$$a_i a_j + b_i b_j cos\,\theta_{ij} = \delta_{ij} \ , \qquad (8.12)$$

θ_{ij} being the angle between the directions of the hybrids. It is to be noted that by using positive orbital coefficients (as defined above), the orthogonality can be achieved only if $\theta_{ij} \geq 90°$. Therefore, the larger is θ_{ij} the more negative is $cos\,\theta_{ij}$. When one increases the angle θ_{ij} then, according to Eq. (8.12), the conservation of orthogonality requires the product $a_i a_j$ to be increased with respect to the product $b_i b_j$. This shows that there is a close connection between the values of the hybridization coefficients a_i, b_i and the valence angles. We shall look for such values of the latter that lead to a molecular geometry with a minimal energy; for that reason we shall first of all take into account that the atomic s-orbitals have lower energies than the p-orbitals.

If one neglects interatomic overlap (e.g., works in a Löwdin-orthogonalized basis), then the overall valence electron population of the central atom is the sum of the populations

$$q_s = 2\sum_i c_i^2 a_i^2 \ , \tag{8.13}$$

and

$$q_p = 2\sum_i c_i^2 b_i^2 \ , \tag{8.14}$$

of the s- and p-orbitals, respectively. As the coefficients c_i are considered approximately constant, the values of q_s and q_p will be determined by the hybridization coefficients a_i, b_i.

As the energies of the s orbitals are lower than those of the p-orbitals, one can achieve an energy decrease if q_s is increased at the expense of q_p. The hybridization transformation (8.11) is *unitary*, from which it follows that

$$\sum_i a_i^2 = 1 \ , \tag{8.15}$$

and this restriction is crucial in understanding the VSEPR rules. In fact, it follows from Eq. (8.15) that the s-population of the central atom may be higher, and thus the energy lower, if the hybrids having higher MO coefficients c_i have increased s-characters, i.e., larger hybridization coefficients a_i. At the same time, the orbitals with smaller c_is should have smaller a_is, in order to keep the sum of squares of a_is equal to one, as required by Eq. (8.15). So, it is energetically advantageous if the largest s-character belongs to the lone pairs ($c_i = 1$), while we have to assign smaller a_i values to the bonding orbitals ($c_i < 1$). The distribution of the s-characters influences the valence angles through the orthogonality relation (8.12) in the following manner:

For an atom with two lone pairs $l = h_i$ and $l' = h_j$, sufficiently large values can be achieved for both coefficients a_i and a_j only if the angle θ_{ij} is opening enough in order to provide $cos\,\theta_{ij}$ to be a sufficiently large negative value that can compensate the decreased values of b_i and b_j. (That decrease follows from the normalization of each hybrid.) Thus the angle between the lone pairs should be the largest—bonding orbitals have smaller a_i values. Accordingly, the orthogonality between a lone pair orbital and a bonding one will be achieved at a smaller angle than that between two lone pairs, and the angle between two bonding orbitals will be even smaller. Therefore, the inequality (8.9) is a direct consequence of the inequality

$$a_l a_l' > a_l a_b > a_b a_b' \ , \tag{8.16}$$

which, in turn, follows from the energy difference between atomic s- and p-levels

If the localized MO φ_i is directed to a ligand that is more electronegative than the central atom, then there will be a shift of electrons toward the ligand, resulting in a reduced value of the respective c_i coefficient. That also reduces the impact of such an orbital on determining the atomic s-population. Therefore, some further energy gain may be achieved if the s-character of such an orbital is reduced, permitting one to increase the s-character of the lone pairs and unpolarized bonding orbitals. The reduced a_i value results in smaller valence angles θ_{ij}, explaining the second rule, according to which the space requirement of the bonds directed to electronegative atoms is smaller than that of the unpolarized bonds. (Of course, the opposite effect will be observed for bonds in which the central atom is the more electronegative partner.) Thus we have given a theoretical explanation to Bent's empirical rules based on the analysis of NMR chemical shifts, etc. According to Bent, "atomic s-character concentrates in orbitals directed toward electropositive substituents," and "atomic p-character concentrates in orbitals directed toward electronegative substituents" [107].

8.4 The CECA method and related schemes

8.4.1 The problem of multicenter integrals: The projective integral expansion scheme

The total energy of a single determinant wave function can be written (admitting UHF case) as [5]

$$\begin{aligned} E = &\sum_{i=1}^{N_\alpha}\langle a_i|\hat{h}|a_i\rangle + \sum_{i=1}^{N_\beta}\langle b_i|\hat{h}|b_i\rangle + \frac{1}{2}\sum_{i,j=1}^{N_\alpha}\left([a_ia_j|a_ia_j] - [a_ia_j|a_ja_i]\right) \\ &+ \frac{1}{2}\sum_{i,j=1}^{N_\beta}\left([b_ib_j|b_ib_j] - [b_ib_j|b_jb_i]\right) + \sum_{i=1}^{N_\alpha}\sum_{j=1}^{N_\beta}[a_ib_j|a_ib_j] + \sum_{A<B}\frac{Z_AZ_B}{R_{AB}}, \end{aligned} \tag{8.17}$$

where a_i and b_i are the molecular orbitals occupied by spins α and β, respectively, $[a_ia_j|a_ia_j]$, etc., are the two-electron integrals over the MOs in the $[12|12]$ convention, Z_A is the nuclear charge of atom A, and R_{AB} is the distance between the nuclei A and B.* The one-electron part $\hat{h}$ of the Hamil-

*Part of this section is reprinted from my paper [111] with permission from Elsevier.

tonian contains the operator of the kinetic energy and the nuclear attraction to all the nuclei:

$$\hat{h} = -\tfrac{1}{2}\Delta - \sum_A \frac{Z_A}{r_A}\,, \tag{8.18}$$

where $\Delta \equiv \nabla^2$ is the Laplacian and r_A is the distance of the electron from the nucleus A. The MOs are linear combinations of the basis orbitals χ_μ, centered on the individual atoms:

$$a_i = \sum_\mu c_\mu^{\alpha i} \chi_\mu\,; \qquad b_i = \sum_\mu c_\mu^{\beta i} \chi_\mu\,. \tag{8.19}$$

By introducing the usual density matrices $\mathbf{P}^\alpha$, $\mathbf{P}^\beta$, and $\mathbf{D}$ according to equations (3.34)–(3.36), Eq. (8.17) can be rewritten as

$$\begin{aligned} E &= \sum_{\mu,\nu} D_{\mu\nu} \langle \nu | -\tfrac{1}{2}\Delta | \mu \rangle - \sum_A \sum_{\mu,\nu} D_{\mu\nu} \langle \nu | \frac{Z_A}{r_A} | \mu \rangle \\ &+ \tfrac{1}{2} \sum_{\mu,\nu,\rho,\tau} \left(D_{\rho\mu} D_{\tau\nu} - P^\alpha_{\rho\nu} P^\alpha_{\tau\mu} - P^\beta_{\rho\nu} P^\beta_{\tau\mu} \right) [\mu\nu|\tau\rho] + \sum_{A<B} \frac{Z_A Z_B}{R_{AB}}\,. \end{aligned} \tag{8.20}$$

Now, if atom-centered basis sets are used, it may be natural to systematize these terms according to the centers to which the individual basis orbitals belong. The first sum with kinetic energy contains one- and two-center terms only, depending on whether the basis orbitals χ_μ and χ_ν are centered on the same or on different atoms. However, the nuclear attraction terms in the second sum may be also of three-center character: if orbitals χ_μ and χ_ν are centered on different atoms, and neither of them coincides with atom A the interaction with the nucleus of which is actually considered. The two-electron part, of course, contains terms of up to four-center: the four orbitals may all be centered on different atoms. As already noted on page 21, a decomposition on such an assignment had been studied by Clementi [21] (as early as 1967) and it led to three- and four-center terms that were too large for the results to be chemically useful.

No such problem appears in the semiempirical theories which use one- and two-center integrals only. Following the work of Fischer and Kollmar [61] on CNDO, the decomposition of the energy into one- and two-center contributions, corresponding very well to the chemist's way of thinking, was implemented for different semiempirical schemes. The results proved to be of significant interpretative power, and sometimes could even be used for predictions. For instance, the use of energy decomposition in the framework of the MNDO theory [75]—combined with the "quasi-Koopmans" scheme already mentioned in Section 6.9—was applied successfully for predicting the primary cleavages in different fields of mass spectrometry [75,79].

The success of the energy decomposition at the semiempirical level motivated us to approach the *ab initio* case in the following manner: one should analyze the different three- and four-center contributions and compress them into one- and two-center ones as far as possible by performing appropriate projections. (The method is essentially the same as that applied earlier in the so-called "chemical Hamiltonian approach" [26].) The idea of these projections can be described as follows.

Let us first consider the case of the three-center integral $\langle\chi_\nu^C|\frac{Z_A}{r_A}|\chi_\mu^B\rangle$, where the superscripts B and C indicate that the orbitals χ_μ and χ_ν are centered on the atoms B, and C, respectively.* (All three atoms A, B, and C are assumed different.) This integral may be considered as the overlap integral of two functions: the atomic orbital χ_ν^C in the "bra" and the function $\frac{Z_A}{r_A}|\chi_\mu^B$ in the "ket." This latter function has an obvious diatomic character: pictorially speaking it describes how the electron on orbital χ_μ^B is "scattered" by the nucleus of atom A. (There is no real scattering process of this type, of course, in a molecule.) This function can be considered as consisting of two components: one can be expanded as a linear combination of the basis function centered on the given pair of atoms A and B, and another which is orthogonal to the subspace of these orbitals. This fact can be written as the identity

$$\frac{Z_A}{r_A}|\chi_\mu^B\rangle \equiv \hat{P}_{AB}\frac{Z_A}{r_A}|\chi_\mu^B\rangle + \left(1-\hat{P}_{AB}\right)\frac{Z_A}{r_A}|\chi_\mu^B\rangle\,, \tag{8.21}$$

where $\hat{P}_{AB}$ is the projector on the subspace representing the union of the subspaces of the basis orbitals centered on atoms A and B, respectively. As the basis orbitals are, in general, not orthogonal to each other, the projector is defined, in full analogy with that in Eq. (2.1), as

$$\hat{P}_{AB} = \sum_{\mu,\nu\in AB} |\chi_\mu\rangle S^{-1}_{(AB)\mu\nu}\langle\chi_\nu|\,, \tag{8.22}$$

where AB indicates the union of the two atomic subspaces and the notation $\in AB$ means that the summation is over the basis orbitals belonging to that

---

*Note that in derivations of this type one has to distinguish between integrals $\langle\chi_\nu^C|\frac{Z_A}{r_A}|\chi_\mu^B\rangle$ and $\langle\chi_\mu^B|\frac{Z_A}{r_A}|\chi_\nu^C\rangle$, that are equal if real orbitals are used. This can be best achieved if in the derivations one consistently admits the use of complex orbitals (and orbital coefficients). As noted in [5], this practice can be recommended in all theoretical studies.

union. Accordingly, $\mathbf{S}_{(AB)}^{-1}$ is the inverse overlap matrix for that diatomic basis set. The first, "diatomic," component on the right-hand side of Eq. (8.21) is always present when the given pair of atoms carrying a given basis set is there at the given distance, while it depends on random factors (the position and basis sets of the other atoms) whether and to what extent the second component actually manifests in the calculations.* It is, therefore, meaningful to *replace* (approximately) the function $\frac{Z_A}{r_A}|\chi_\mu^B\rangle$ with its diatomic component:

$$\frac{Z_A}{r_A}|\chi_\mu^B\rangle \Longrightarrow \hat{P}_{AB}\frac{Z_A}{r_A}|\chi_\mu^B\rangle \,. \tag{8.23}$$

Then the integral $\langle\chi_\nu^C|\frac{Z_A}{r_A}|\chi_\mu^B\rangle$ we are considering will be replaced by:

$$\langle\chi_\nu^C|\frac{Z_A}{r_A}|\chi_\mu^B\rangle \Longrightarrow \langle\chi_\nu^C|\hat{P}_{AB}\frac{Z_A}{r_A}|\chi_\mu^B\rangle = \sum_{\rho,\tau\in AB} S_{\nu\rho}S_{(AB)\rho\tau}^{-1}\langle\chi_\tau^{AB}|\frac{Z_A}{r_A}|\chi_\mu^B\rangle \,. \tag{8.24}$$

Here the superscript "AB" stresses that orbital χ_τ^{AB} is centered either on atom A or atom B; therefore the right-hand side of this (approximate) expansion does not contain three-center nuclear attraction integrals any more, only one- and two-center ones.

Of course, one could apply the above expansion also in the case when atom C coincides with either A or B; it is easy to see that then Eq. (8.24) does not represent an approximation, but an identity: in that case $\nu\in AB$, too, so $\sum_{\rho\in AB} S_{\nu\rho}S_{(AB)\rho\tau}^{-1} = \delta_{\nu\rho}$. Otherwise the expansion (8.24) would be exact only in the limiting case when the AB basis was complete. In real cases the error of this approximation (the difference of the two sides) represents a sort of a—usually small—"white noise"; we shall see that there is a way to account for this "noise" if required. One can generalize the above scheme also for the case of two-electron integrals, as will be described in detail in the next section.

The projective integral approximation scheme discussed above is utilized in the "chemical Hamiltonian approach" (CHA)—as applied to both the intramolecular effects [26] and to treat the so-called "basis set superposition error" (BSSE) problem of intermolecular interactions (e.g., [108,109])—and serves the basis for the energy decomposition scheme called "chemical energy component analysis" (CECA), which we are going to discuss below. An important, although somewhat formal, difference between the CHA treatment

*The term containing the projector $1-\hat{P}_{AB}$ will manifest if it is not orthogonal to some basis functions outside the diatomic fragment AB considered.

and CECA is connected with the fact that the expressions like Eq. (8.24) are asymmetric with respect to "bra" and "ket" parts of the integral. In CHA this asymmetry is conserved, and leads to a non-Hermiticity of the CHA Hamiltonians* [26, 108], while in CECA all the expressions are symmetrized. Thus, instead of Eq. (8.24), the symmetrized formula

$$\langle\chi_\nu^C|\frac{Z_A}{r_A}|\chi_\mu^B\rangle \Longrightarrow \frac{1}{2}\Bigg\{\sum_{\rho,\tau\in AB} S_{\nu\rho}S^{-1}_{(AB)\rho\tau}\langle\chi_\tau^{AB}|\frac{Z_A}{r_A}|\chi_\mu^B\rangle + \Bigg(\sum_{\rho,\tau\in AC} S_{\mu\rho}S^{-1}_{(AC)\rho\tau}\langle\chi_\tau^{AC}|\frac{Z_A}{r_A}|\chi_\nu^C\rangle\Bigg)^*\Bigg\}, \tag{8.25}$$

is used. It is to be noted that for molecules the importance of these symmetrizations is rather conceptual than practical, as one usually applies real basis orbitals and orbital coefficients, and the density matrix is symmetric. However, the CECA scheme had also been applied [110] to infinite periodic systems in which the Bloch-orbitals are inherently complex, so there the symmetrization is more than a pure theoretical exercise.

Soon after the CECA method was introduced [111], it was realized [112] that the Hartree–Fock energy of a molecule "spontaneously" decomposes into a sum of atomic and diatomic energy components in the framework of Bader's 3D analysis (see Section 9.6), and a "mapping" has also been found permitting to obtain the CECA energy expressions from their 3D counterparts, supporting the CECA idea to a great extent.

8.4.2 The integrals in the CECA method

In CECA the kinetic energy operator is always combined with an electron-nuclear attraction one to form the intraatomic one-electron Hamiltonian:

$$\hat{h}^A = -\frac{1}{2}\Delta - \frac{Z_A}{r_A}, \tag{8.26}$$

so the (one- or two-center) integrals of type Eq. (8.24) with $A = B$ are treated in this framework. One again should have a symmetric treatment, that can be

*The CHA treatment of the intramolecular interactions permitted to write down the Born–Oppenheimer Hamiltonian applied to the Hilbert-space treatment using atom-centered (finite) basis sets, as a sum of atomic and diatomic terms and some finite basis corrections [26]. In the intermolecular domain it permitted to identify and omit those terms of the Hamiltonian that cause BSSE. Using the BSSE-free Hamiltonian one can obtain wave functions that are not spoiled by BSSE effects; then the energy should be calculated as a conventional energy expectation value over these BSSE-free wave functions by using the full original Hamiltonian (so-called "CHA/CE scheme") [108].

achieved by writing for the matrix element of the whole one-electron Hamiltonian [111] as

$$\langle\chi_\nu^B|\hat{h}|\chi_\mu^A\rangle = \frac{1}{2}\Bigg\{\langle\chi_\nu^B|\hat{h}^A|\chi_\mu^A\rangle - \sum_{C\neq A}\langle\chi_\nu^B|\frac{Z_C}{r_C}|\chi_\mu^A\rangle + \left(\langle\chi_\mu^A|\hat{h}^B|\chi_\nu^B\rangle - \sum_{C\neq B}\langle\chi_\mu^A|\frac{Z_C}{r_C}|\chi_\nu^B\rangle\right)^*\Bigg\}. \tag{8.27}$$

The three-center nuclear-attraction integrals in this expression should be treated by the approximation (8.24)—the symmetrization (8.25) is explicitly included in Eq. (8.27).

Inserting a resolution of identity $1 \equiv \hat{P}_A + (1-\hat{P}_A)$, where

$$\hat{P}_A = \sum_{\mu,\nu\in A} |\chi_\mu\rangle S^{-1}_{(A)\mu\nu}\langle\chi_\nu|, \tag{8.28}$$

is the projector on the basis orbitals of atom A, into the integral $\langle\chi_\nu^B|\hat{h}^A|\chi_\mu^A\rangle$, one can separate out a leading component containing only intraatomic integrals of the operator $\hat{h}^A$, and the remainder:

$$\begin{aligned}\langle\chi_\nu^B|\hat{h}^A|\chi_\mu^A\rangle &= \langle\chi_\nu^B|\hat{P}_A\hat{h}^A|\chi_\mu^A\rangle + \langle\chi_\nu^B|(1-\hat{P}_A)\hat{h}^A|\chi_\mu^A\rangle \\ &= \sum_{\rho,\tau\in A} S_{\nu\rho}S^{-1}_{(A)\rho\tau}\langle\chi_\tau^A|\hat{h}^A|\chi_\mu^A\rangle + \langle\chi_\nu^B|(1-\hat{P}_A)\hat{h}^A|\chi_\mu^A\rangle .\end{aligned} \tag{8.29}$$

The remainder containing the operator $1-\hat{P}_A$, obviously, vanishes in the special case when $B = A$ and we are dealing with a one-center integral which needs no special treatment. (In that case $\langle\chi_\nu^A|(1-\hat{P}_A) = 0$, owing to the Hermiticity of the projectors.) In the diatomic case ($B \neq A$) representing the real interest, the remainder can be calculated as the difference of the original integral in the left-hand side and the sum on the right-hand side; while the sum contains only one-center one-electron integrals, the remainder is inherently diatomic.

The two-electron integrals are treated in an essentially similar manner. The "ket" function

$$\frac{1}{r_{12}}|\chi_\mu^A(1)\chi_\nu^B(2)\rangle, \tag{8.30}$$

is considered to be of one- or diatomic nature, depending on whether $A = B$ or $A \neq B$, and is projected on the subspace of the atomic or diatomic set of basis orbitals by using two projectors: one for electron "1," another for electron "2." We perform a symmetrization, similar to that considered in the one-electron case:

$$\langle\chi_\rho^C(1)\chi_\tau^D(2)|\frac{1}{r_{12}}|\chi_\mu^A(1)\chi_\nu^B(2)\rangle=\frac{1}{2}\left\{\langle\chi_\rho^C(1)\chi_\tau^D(2)|\frac{1}{r_{12}}|\chi_\mu^A(1)\chi_\nu^B(2)\rangle\right.$$
$$\left.+\langle\chi_\mu^A(1)\chi_\nu^B(2)|\frac{1}{r_{12}}|\chi_\rho^C(1)\chi_\tau^D(2)\rangle^*\right\}. \quad (8.31)$$

The integrals on the right-hand side are then replaced by their projective approximations. In the special case, when $A = B$ in the "ket," one has

$$\langle\chi_\rho^C(1)\chi_\tau^D(2)|\frac{1}{r_{12}}|\chi_\mu^A(1)\chi_\nu^A(2)\rangle\Longrightarrow\langle\chi_\rho^C(1)\chi_\tau^D(2)|\hat{P}_A(1)\hat{P}_A(2)\frac{1}{r_{12}}\chi_\mu^A(1)\chi_\nu^A(2)\rangle$$
$$=\sum_{\kappa,\vartheta,\lambda,\sigma\in A}S_{\rho\kappa}S^{-1}_{(A)\kappa\vartheta}S_{\tau\lambda}S^{-1}_{(A)\lambda\sigma}\langle\chi_\vartheta^A(1)\chi_\sigma^A(2)|\frac{1}{r_{12}}|\chi_\mu^A(1)\chi_\nu^A(2)\rangle. \quad (8.32)$$

The remainder of this approximation is treated depending on whether or not $C = D$. In the first case the initial integral is a diatomic one, then—similar to the integrals of $\hat{h}^A$—the remainder is considered of two-center nature, and it will contribute to the respective diatomic energy component. If $C \neq D$ the remainder is of three-center character and is omitted in this analysis.

If $A \neq B$, then the function $\frac{1}{r_{12}}|\chi_\mu^A(1)\chi_\nu^B(2)\rangle$ is of diatomic nature and we use the diatomic projector $\hat{P}_{AB}$. Accordingly, we shall write

$$\langle\chi_\rho^C(1)\chi_\tau^D(2)|\frac{1}{r_{12}}|\chi_\mu^A(1)\chi_\nu^B(2)\rangle\Longrightarrow \quad (8.33)$$
$$\Longrightarrow\langle\chi_\rho^C(1)\chi_\tau^D(2)|\hat{P}_{AB}(1)\hat{P}_{AB}(2)\frac{1}{r_{12}}\chi_\mu^A(1)\chi_\nu^B(2)\rangle.$$

Obviously, this expression becomes an exact equality if the set of atoms $\{C,D\}$ is a (sub)set of $\{A,B\}$; then one has a two-center integral which will be treated without introducing any approximations. If C and/or D really represent a third (fourth) atom, then the remainder of the approximation (8.33) is an inherently three- or four-atom correction and is omitted in the further considerations. Expanding the projectors in (8.33) we get the explicit expansion of this integral as

$$\langle\chi_\rho^C(1)\chi_\tau^D(2)|\frac{1}{r_{12}}|\chi_\mu^A(1)\chi_\nu^B(2)\rangle\Longrightarrow \quad (8.34)$$
$$\Longrightarrow\sum_{\kappa,\vartheta,\lambda,\sigma\in AB}S_{\rho\kappa}S^{-1}_{(AB)\kappa\vartheta}S_{\tau\lambda}S^{-1}_{(AB)\lambda\sigma}\langle\chi_\vartheta^{AB}(1)\chi_\sigma^{AB}(2)|\frac{1}{r_{12}}|\chi_\mu^A(1)\chi_\nu^B(2)\rangle.$$

The superscripts "*A*," "*AB*," etc., of the orbitals were used here to make the derivations more transparent; in the following they will not be applied, as the limitations on the sums like $\mu \in A$ or $\mu \in AB$ will be sufficient to denote the atom(s) to which the given basis orbital belongs.

8.4.3 The CECA energy components

Assuming the orbitals and orbital coefficients to be real and substituting the above projective expansions of the different integrals into the single determinant SCF energy formula (8.20), one gets, after some somewhat lengthy but quite straightforward algebra, the approximate expansion of the molecular energy in terms of one- and two-center components (UHF case is admitted):

$$E \Longrightarrow \sum_A E_A + \sum_{A<B} E_{AB}, \tag{8.35}$$

where

$$E_A = \sum_{\nu,\tau \in A} B^A_{\nu\tau} h^A_{\tau\nu} + \tfrac{1}{2} \sum_{\kappa,\rho,\tau,\eta \in A} [\tau\eta|\kappa\rho] \left(B^A_{\kappa\tau} B^A_{\rho\eta} - C^{\alpha A}_{\kappa\eta} C^{\alpha A}_{\rho\tau} - C^{\beta A}_{\kappa\eta} C^{\beta A}_{\rho\tau} \right), \tag{8.36}$$

and

$$\begin{aligned} E_{AB} = {} & \frac{Z_A Z_B}{R_{AB}} - \sum_{\tau \in AB} \left(\sum_{\mu \in A} B^{AB}_{\mu\tau} < \tau | \frac{Z_B}{r_B} | \mu > + \sum_{\mu \in B} B^{AB}_{\mu\tau} < \tau | \frac{Z_A}{r_A} | \mu > \right) \\ & + \sum_{\substack{\kappa \in A \\ \rho \in B}} \sum_{\tau,\eta \in AB} [\tau\eta|\kappa\rho] \left(B^{AB}_{\kappa\tau} B^{AB}_{\rho\eta} - C^{\alpha AB}_{\kappa\eta} C^{\alpha AB}_{\rho\tau} - C^{\beta AB}_{\kappa\eta} C^{\beta AB}_{\rho\tau} \right) \\ & + \sum_{\substack{\nu \in A \\ \mu \in B}} D_{\mu\nu} \left[h^A_{\mu\nu} - \sum_{\tau \in A} A^A_{\mu\tau} h^A_{\tau\nu} + h^B_{\mu\nu} - \sum_{\tau \in B} A^B_{\nu\tau} h^B_{\tau\mu} \right] \\ & + \tfrac{1}{2} \sum_{\kappa,\rho \in A} \sum_{\substack{\gamma,\nu \in AB \\ (\gamma \notin A) \vee (\nu \notin A)}} \left(D_{\kappa\gamma} D_{\rho\nu} - P^\alpha_{\kappa\nu} P^\alpha_{\rho\gamma} - P^\beta_{\kappa\nu} P^\beta_{\rho\gamma} \right) \\ & \qquad \times \left[[\gamma\nu|\kappa\rho] - \sum_{\tau,\eta \in A} A^A_{\gamma\tau} A^A_{\nu\eta} [\tau\eta|\kappa\rho] \right] \\ & + \tfrac{1}{2} \sum_{\kappa,\rho \in B} \sum_{\substack{\gamma,\nu \in AB \\ (\gamma \notin B) \vee (\nu \notin B)}} \left(D_{\kappa\gamma} D_{\rho\nu} - P^\alpha_{\kappa\nu} P^\alpha_{\rho\gamma} - P^\beta_{\kappa\nu} P^\beta_{\rho\gamma} \right) \\ & \qquad \times \left[[\gamma\nu|\kappa\rho] - \sum_{\tau,\eta \in B} A^B_{\gamma\tau} A^B_{\nu\eta} [\tau\eta|\kappa\rho] \right]. \end{aligned} \tag{8.37}$$

We note that in the original paper [111] the convention $(11|22)$ was used for the two-electron integrals, which is usually more convenient in the programming work. Here, however, we stick to the $[12|12]$ convention more adequate for theoretical considerations. Furthermore, in the definitions of the different auxiliary matrices we shall consider X as a notation for a subunit—either atomic (e.g., $X = A$) or diatomic ($X = AB$). Matrix $\mathbf{A}^X$ is closely related to the projector $\hat{P}_X$ and is defined as [113]

$$A^X_{\lambda\mu} = \sum_{\sigma\in X} S_{\lambda\sigma} S^{-1}_{(X)\sigma\mu} \qquad (\mu \in X)\,. \tag{8.38}$$

As is easy to see, the intrafragment block of each matrix $\mathbf{A}^X$ is a unit matrix: $A^X_{\lambda\mu} = \delta_{\lambda\mu}$, if $\lambda, \mu \in X$.

Matrices $\mathbf{B}^X$ and $\mathbf{C}^{\sigma X}$ represent the combination of the density matrices and of the projection on the subunit X:

$$\begin{aligned} B^X_{\mu\nu} &= \sum_\gamma D_{\mu\gamma} A^X_{\gamma\nu} && (\nu \in X) \\ C^{\sigma X}_{\mu\nu} &= \sum_\gamma P^\sigma_{\mu\gamma} A^X_{\gamma\nu} && (\sigma = \alpha \text{ or } \beta;\;\; \nu \in X) \end{aligned} \tag{8.39}$$

As is easy to see, the energy components E_A, E_{AB} contain terms of two types. One of them resembles a term of the original energy formula (8.20), but the density matrices $\mathbf{D}$ and $\mathbf{P}^\alpha$, $\mathbf{P}^\beta$ are replaced by their "projected counterparts" $\mathbf{B}^X$, $\mathbf{C}^{\alpha X}$, $\mathbf{C}^{\beta X}$, respectively. (E_A contains terms of this type only.) Another type of term contains the difference between some two-center integrals and their one-center approximations, given by the square brackets in Eq. (8.37).* As noted on page 140, the aim of the projections is to reduce the different interactions to one- and two-center effects, as far as possible. In accord with that, no three- or four-center one- and two-electron integrals appear explicitly in the formulae for E_A and E_{AB}. (The three- and four-center effects manifest through the respective "projected density matrices" $\mathbf{B}^X$, $\mathbf{C}^{\sigma X}$.) Intraatomic terms are analogously projected on the atomic subspaces; however the differences between the two-center integrals and their one-center approximations are conserved and contribute to the diatomic energy component—these are the terms with the square brackets mentioned above. In this manner we could achieve that *the CECA energy components sum to the exact total energy in the case of diatomic molecules.*

*In accord with this, the restrictions like $(\gamma \notin A) \vee (\nu \notin A)$ on the last sums, indicating that at least one of the orbitals χ_γ or χ_ν is centered outside atom A, may be omitted, as in the case of $\gamma, \nu \in A$ we have a one-center integral, for which the respective difference vanishes.

The accuracy of the integral approximations introduced may be guessed on the basis of comparing the exact SCF energies and the sum of the CECA one- and two-center energy components of a given molecule. In Ref. 111 such a comparison was done for the ethane molecule, by using a wide variety of basis sets from 6-31G to 6-311++G** and cc-pVDZ, and it was found that the total energy of about −79.2 Hartrees of this molecule in all cases was approximated within 15 *milli*Hartrees, and the deviation was less than 20 *m*H even for 4-31G. As these deviations are randomly distributed between different parts of the molecule, they indeed represent a "white noise" to which no physical or chemical importance should be attributed. (Nonetheless, if required, they can be accorded a special careful treatment as will be described in the next section.)

Table 8.3
Energy components of the ethane, ethylene, and acetylene molecules calculated by using the 6-31G** basis sets[a]

Atom(s)	Energy component (a.u.)		
	Ethane	Ethylene	Acetylene
C	-36.849	-36.811	-36.712
H	-0.249	-0.241	-0.220
C–C	-0.599	-1.092	-1.761
C–H	-0.605	-0.611	-0.564
H–H geminal	0.016	0.012	–
H–H vicinal	0.0036 (2x)	0.0045	0.0059
	-0.0007 (1x)	0.0003	
C–H vicinal	0.0097	0.0124	-0.0243

[a]The total SCF energies/sums of the one- and two-center components are -79.238235/-79.2489290, -78.038841/-78.027549 and -76.821837/-76.794210 a.u., respectively.

*[111] with permission from Elsevier.

Tables 8.3 to 8.5 contain some examples of the one-center and two-center energy components. The following general tendency can be observed. The two-center energy components are too negative (too large in absolute values) compared with the typical dissociation energies of the chemical bonds. Accordingly, the one-center components are shifted to the positive direction, i.e., are less negative than one would expect. Interestingly enough, the same phenomenon appeared also at the semiempirical level (say MNDO) of theory, where energy components were often used. The problem that the energy components one obtains by different straightforward semiempirical and *ab initio* energy decomposition schemes "are not on the chemical scale" motivated a number of investigations (c.f. Section 8.4.5); our solution of it will be considered in Section 8.5. Disregarding the fact that the CECA diatomic energy components are too large in absolute values, they are quite adequate to *compare* similar bonds in different molecules; of course, only numbers obtained by using strictly the same basis set should be compared.

Table 8.4
Energy components of the benzene molecule calculated by using the 6-31G** basis sets[a]

Atom(s)	Energy component (a.u.)
C	-36.722
H	-0.233
C_1–C_2	-0.939
C_1–C_3	0.0301
C_1–C_4	0.0086
C_1–H_7	-0.603
C_1–H_8	0.0086
C_1–H_9	-0.0039
C_1–H_{10}	-0.0024
H_7–H_8	0.0045
H_7–H_9	0.0022
H_7–H_{10}	0.0019

[a]The total SCF energy and the sum of the one-and two-center components are -230.713860 and -230.686564 a.u., respectively.
*[111] with permission from Elsevier.

Perhaps 4-31G is the smallest basis set for which the accuracy of the CECA integral approximations is sufficient to get an insight into the most important interactions in a molecular system: one can not only identify the chemically bonded atoms giving large negative diatomic contributions or the repulsive non-bonded interactions of closed shells (much smaller positive diatomic components) but also the intra- or inter-molecular hydrogen bonds, and even different subtle effects like hydrogen-bond-like interactions (e.g., [73,114]). In our practice the use of the sufficiently "atom-like," but polarized 6-31G** —in other notations 6-31G(d,p)—basis set has been found very useful.

The fact that the total energy can be presented to a good accuracy as a sum of atomic and (bonding or non-bonded) diatomic energy components shows that the notion of predominantly intraatomic and pairwise interatomic interactions, as implied by the traditional chemical formulae, agrees very well with the *ab initio* results. (Of course, every atomic and interatomic energy component strongly depends on the chemical environment.) In this aspect, the picture is analogous to the picture obtained with the use of bond order indices. However, as noted on page 75, the bond orders of chemically non-bonded pairs of atoms not always can distinguish between situations with attractive, repulsive, and neutral resulting interactions, as atoms in repulsive situations can also exhibit positive bond orders; energy components are expected to be much more useful in such cases.

Table 8.5
Energy components of the diborane molecule calculated by using the 6-31G** basis sets[a]

Atom(s)	Energy component (a.u.)
B	-23.826
H bridge	-0.299
H terminal	-0.318
B–H bridge	-0.279
B–H terminal	-0.517
B–B	-0.227
H bridge–H bridge	0.0253
H bridge–H terminal	0.0044
H terminal–H terminal	0.0057
H terminal–H terminal′	0.0011
H terminal–H terminal″	-0.0005

[a]The total SCF energy and the sum of the one- and two-center components are -52.819860 and -52.838029 a.u., respectively.
*[111] with permission from Elsevier.

It is to be noted that the use of atomic and diatomic energy components only does not mean that this scheme is unable to account for collective effects. In this framework collective effects manifest explicitly in the wave functions, but through that they influence the actual values of the energy components. Thus, for instance, the B–B energy component corresponding to the "half" B–B bond created by the two-electron three-center interactions is roughly the same as the B–H_{bridge} energy component, which also corresponds to a bond order of about 1/2.

The energy components obtained in the CECA analysis may not only used for interpretative purposes, but are as good—sometimes even better—for predicting mass spectra, as are the bond orders, the use of which was discussed in Section 6.9. Our free programs Ref. 77 and 78 perform the calculation of both bond orders and energy components at the MNDO and *ab initio* levels, respectively.

8.4.4 The three- and four-center corrections

As discussed above, the CECA energy components provide an *approximate* decomposition of the total SCF energy into one- and two-center components. Accordingly, their sum is not *strictly* equal to the exact SCF energy of the molecule. The deviations lack any systematic character and can hardly be attributed any physical meaning. These essentially random numbers ("white noise") may be, in fact, simply neglected. However, one might also feel it desirable to distribute these small effects between the individual one- and two-center energy components in some systematic manner, so as to get energy components that exactly sum up to the total molecular energy. Now we are going to describe an appropriate scheme that permits to achieve this goal.

We shall consider an extremely simple alternative energy decomposition scheme [100,115], in which every term assigned—without performing any projections—to a respective atom or pair of atoms simply by considering the basis functions in the "ket" part of the integral it is containing.* This means that the terms with integrals like $\langle \chi_\nu^C | \frac{Z_A}{r_A} | \chi_\mu^B \rangle$ and $\langle \chi_\rho^C(1)\chi_\tau^D(2) | \frac{1}{r_{12}} | \chi_\mu^A(1)\chi_\nu^B(2) \rangle$ are simply considered as diatomic ones, corresponding to the pair of atoms A and B (if $A \neq B$) or as atomic ones, corresponding to atom A (if $A = B$), irrespective of whether or not the atoms C and D coincide with either of A or B.† In this manner the energy spontaneously decomposes into a sum of terms of atomic or diatomic character, and one

*That is equivalent to inserting one and two "atomic resolutions of identity (3.1) with the definition (3.8) to the "ket" part of each one- and two-electron integral in the energy formula.

†We shall recall here the remark made in the footnote on page 140.

obtains an exact decomposition of the energy

$$E = \sum_A E_A + \sum_{A<B} E_{AB} \tag{8.40}$$

with the components

$$E_A^{E1} = \sum_{\mu\in A}\sum_{\nu} D_{\mu\nu}h_{\nu\mu}^A + \frac{1}{2}\sum_{\mu,\rho\in A}\sum_{\nu,\tau}(D_{\mu\nu}D_{\rho\tau} - P_{\rho\nu}^{\alpha}P_{\mu\tau}^{\alpha} - P_{\rho\nu}^{\beta}P_{\mu\tau}^{\beta})[\nu\tau|\mu\rho] \tag{8.41}$$

and

$$E_{AB}^{E1} = \frac{Z_A Z_B}{R_{AB}} - \sum_{\mu\in A}\sum_{\nu} D_{\mu\nu}\langle\chi_\nu|\frac{Z_B}{r_B}|\chi_\mu\rangle - \sum_{\mu\in B}\sum_{\nu} D_{\mu\nu}\langle\chi_\nu|\frac{Z_A}{r_A}|\chi_\mu\rangle + \sum_{\mu\in A}\sum_{\rho\in B}\sum_{\nu,\tau}(D_{\mu\nu}D_{\rho\tau} - P_{\rho\nu}^{\alpha}P_{\mu\tau}^{\alpha} - P_{\rho\nu}^{\beta}P_{\mu\tau}^{\beta})[\nu\tau|\mu\rho] \tag{8.42}$$

Here we have introduced the superscript $E1$ to indicate that these components correspond to an exact decomposition (8.40) and to distinguish it from another related scheme which will be denoted by $E2$ in Section 8.4.5.

Now, one can introduce *the same* projective integral approximations as were applied in the CECA scheme, and get an analogous approximation of total energy

$$E \Longrightarrow \sum_A E_A^{A1} + \sum_{A<B} E_{AB}^{A1} , \tag{8.43}$$

where the superscript A1 indicates that we are dealing with an approximations to the scheme E1.

As is easy to see, in the atomic components Eq. (8.41) the one-electron integrals are either one-center or two-center ones, for which no approximations need to be introduced. The two-electron integrals, however, are up to three-center ones. (As $\mu, \rho \in A$, no four-center integrals occur in this expression.) According to Eq. (8.32) and definition (8.38), these three-center integrals are approximated in CECA as:

$$[\nu\tau|\mu\rho] \Longrightarrow \sum_{\eta,\kappa} A_{\nu\eta}^A A_{\tau\kappa}^A [\eta\kappa|\mu\rho] \,. \tag{8.44}$$

As we wish to find some corrections to the CECA scheme, we have to separate out the corresponding one-center component of CECA. After adding and subtracting some terms, one gets:

$$E_A^{A1} = \sum_{\mu,\tau\in A} B_{\mu\tau}^A h_{\tau\mu}^A + \frac{1}{2}\sum_{\mu,\nu,\rho,\tau\in A}(B_{\mu\nu}^A B_{\rho\tau}^A - C_{\rho\nu}^{A\alpha}C_{\mu\tau}^{A\alpha} - C_{\rho\nu}^{A\beta}C_{\mu\tau}^{A\beta})[\nu\tau|\mu\rho]$$

$$+\sum_{\substack{B\\B\neq A}}\left\{\sum_{\mu\in A}\sum_{\nu\in B}D_{\mu\nu}\left[h^{A}_{\nu\mu}-\sum_{\tau\in A}A^{A}_{\nu\tau}h^{A}_{\tau\mu}\right]\right. \tag{8.45}$$

$$+\frac{1}{2}\sum_{\mu,\rho\in A}\sum_{\substack{\nu,\tau\in AB\\(\nu\notin A)\vee(\tau\notin A)}}(D_{\mu\nu}D_{\rho\tau}-P^{\alpha}_{\rho\nu}P^{\alpha}_{\mu\tau}-P^{\beta}_{\rho\nu}P^{\beta}_{\mu\tau})$$

$$\left.\times\left[[\nu\tau|\mu\rho]-\sum_{\eta,\kappa\in A}A^{A}_{\nu\eta}A^{A}_{\tau\kappa}[\eta\kappa|\mu\rho]\right]\right\}$$

Here the first terms reproduce the CECA one-center components, while the remaining ones contain the differences between some two-electron integrals and their one-center approximations. These terms are nothing more than those terms of the CECA two-center energy components that contain the square braces, only here they are assigned to the one-center components and not to the two-center ones.

Using the projective approximations (8.25) and (8.31) in the two-center component (8.42), we get the approximate expression

$$E^{A1}_{AB}=\frac{Z_AZ_B}{R_{AB}}-\sum_{\mu\in A}\sum_{\tau\in AB}B^{AB}_{\mu\tau}\langle\chi_\tau|\frac{Z_B}{r_B}|\chi_\mu\rangle-\sum_{\mu\in B}\sum_{\tau\in AB}B^{AB}_{\mu\tau}\langle\chi_\tau|\frac{Z_A}{r_A}|\chi_\mu\rangle$$
$$+\sum_{\mu\in A}\sum_{\rho\in B}\sum_{\nu,\tau\in AB}(B^{AB}_{\mu\nu}B^{AB}_{\rho\tau}-C^{AB\alpha}_{\rho\nu}C^{AB\alpha}_{\mu\tau}-C^{AB\beta}_{\rho\nu}C^{AB\beta}_{\mu\tau})[\nu\tau|\mu\rho] \tag{8.46}$$

This formula reproduces that of CECA, except those terms in square braces that have been regrouped to the one-center ones.

The difference between the energy components obtained with and without introducing the integral approximations (i.e., between the schemes denoted A1 and E1) reflects the remaining minor effects of the three- and four-center terms, that could not be accounted for by the projective integral expansions. Accordingly, these differences may be defined as the three- and four-center corrections to the respective one- and two-center energy components. Now, it is becoming important the fact that the sum of the approximate energy components obtained in scheme A1 is exactly the same as that in the CECA. The approximate energy components of this scheme and of CECA differ only in regrouping some terms between the one- and two-center energy contributions, and *the regrouped terms do not change when approximating the integrals.* Therefore, we may assume that the same three- and four-center corrections apply also to the CECA energy components, and obtain in this manner a modified CECA providing us an exact energy decomposition scheme. According to the above discussion, its formulae (denoted by the superscript "CE") can be obtained as follows:

$$E_A^{CE} = E_A^{CECA} + E_A^{E1} - E_A^{A1} \; ; \tag{8.47}$$

$$E_{AB}^{CE} = E_{AB}^{CECA} + E_{AB}^{E1} - E_{AB}^{A1} \; , \tag{8.48}$$

where we introduced the subscript "CECA" for denoting the CECA energy components (8.36) and (8.37). When summing these energy components, the approximate energy components denoted by subscripts CECA and A1 cancel, as they sum to the same value, thus the components denoted CE will indeed sum to the exact energy (as do the components E1). Their explicit expressions can be given as:

$$E_A^{CE} = E_A^{E1} - \sum_{\substack{B \\ (B \neq A)}} \delta_{AB} \; ; \tag{8.49}$$

$$E_{AB}^{CE} = E_{AB}^{E1} + \delta_{AB} + \delta_{BA} \; , \tag{8.50}$$

where E_A^{E1} and E_{AB}^{E1} are given above in Eqs. (8.41), (8.42), and δ_{AB} is defined as

$$\begin{aligned} \delta_{AB} = & \sum_{\mu \in A} \sum_{\nu \in B} D_{\mu\nu} \left[h_{\nu\mu}^A - \sum_{\tau \in A} A_{\nu\tau}^A h_{\tau\mu}^A \right] \\ & + \frac{1}{2} \sum_{\mu,\rho \in A} \sum_{\substack{\nu,\tau \in AB \\ (\nu \notin A) \vee (\tau \notin A)}} (D_{\mu\nu} D_{\rho\tau} - P_{\rho\nu}^{\alpha} P_{\mu\tau}^{\alpha} - P_{\rho\nu}^{\beta} P_{\mu\tau}^{\beta}) \\ & \times \left[[\nu\tau|\mu\rho] - \sum_{\eta,\kappa \in A} A_{\nu\eta}^A A_{\tau\kappa}^A [\eta\kappa|\mu\rho] \right] . \end{aligned} \tag{8.51}$$

δ_{BA} is obtained by interchanging A and B.

8.4.5 Remarks about kinetic energy

In the above schemes the kinetic energy is treated as a part of the intraatomic Hamiltonian and, accordingly, it basically contributes to the atomic energy components. There is, however, a possibility to treat the kinetic energy similarly to the case used when applying the virial theorem to the energy decomposition problem, i.e., as having both one-center and two-center contributions. This leads to the exact energy decomposition scheme (denoted E2), which differs from E1 by regrouping the two-center kinetic energy components to the diatomic contributions. Thus we get the exact decomposition (8.40) with the components:*

$$E_A^{E2} = \sum_{\mu,\nu\in A} D_{\mu\nu}T_{\nu\mu} - \sum_{\mu\in A}\sum_{\nu} D_{\mu\nu}\langle\chi_\nu|\frac{Z_A}{r_A}|\chi_\mu\rangle$$
$$+\frac{1}{2}\sum_{\mu,\rho\in A}\sum_{\nu,\tau}(D_{\mu\nu}D_{\rho\tau} - P^\alpha_{\rho\nu}P^\alpha_{\mu\tau} - P^\beta_{\rho\nu}P^\beta_{\mu\tau})[\nu\tau|\mu\rho] \tag{8.52}$$

$$E_{AB}^{E2} = 2\sum_{\mu\in A}\sum_{\nu\in B} D_{\mu\nu}T_{\nu\mu} - \sum_{\mu\in A}\sum_{\nu} D_{\mu\nu}\langle\chi_\nu|\frac{Z_B}{r_B}|\chi_\mu\rangle - \sum_{\mu\in B}\sum_{\nu} D_{\mu\nu}\langle\chi_\nu|\frac{Z_A}{r_A}|\chi_\mu\rangle$$
$$+\sum_{\mu\in A}\sum_{\rho\in B}\sum_{\nu,\tau}(D_{\mu\nu}D_{\rho\tau} - P^\alpha_{\rho\nu}P^\alpha_{\mu\tau} - P^\beta_{\rho\nu}P^\beta_{\mu\tau})[\nu\tau|\mu\rho] + \frac{Z_A Z_B}{R_{AB}} \tag{8.53}$$

where $T_{\nu\mu} = \langle\chi_\nu| -\frac{1}{2}\Delta|\chi_\mu\rangle$ is a matrix element of the kinetic energy.

Interestingly enough, the numerical values of the energy components obtained by this extremely simple scheme are very "chemical," if calculated at the equilibrium molecular geometries. However, at other geometries this scheme exhibits a counterintuitive behavior [73]: if one considers a geometry with a stretched chemical bond, then the respective diatomic energy component becomes more negative—and not less negative as one would expect. This is due to the fast decay of the kinetic energy integrals with the distance.† We had to conclude [73] that this property essentially prevents us from using this scheme in studies of practical chemical problems, and we had to look for another way of putting the numbers obtained in the energy decomposition "on the chemical scale." This approach will be described in Section 8.5.

8.4.6 Physical analysis of the diatomic CECA energy components

As noted in Section 8.4.3, the formula (8.36) for one-center energy components resembles the standard Hartree–Fock energy, only the elements of the conventional density matrices **D** and $\mathbf{P}^\sigma$ are replaced by the projected ones $\mathbf{B}^A$ and $\mathbf{C}^{\sigma A}$, respectively. Some deeper discussion of the one-center terms will be given in Section 8.5.

---

*Of course, one could introduce the integral approximations of the CECA scheme in these formulae, too, and get another related approximate decomposition scheme.

†We have tried to find a connection between the very reasonable energy components one obtains right at the equilibrium geometries and the virial theorem which also has its particularly simple form at the equilibrium conformations, but have not succeeded.

The two-center energy terms (8.37) are apparently more complex. As noted in Section 8.4.3, they contain the terms

$$E_{AB}^{bas.ext.} = \sum_{\substack{\nu\in A\\ \mu\in B}} D_{\mu\nu}\left[h^A_{\mu\nu} - \sum_{\tau\in A} A^A_{\mu\tau}h^A_{\tau\nu} + h^B_{\mu\nu} - \sum_{\tau\in B} A^B_{\nu\tau}h^B_{\tau\mu}\right] \tag{8.54}$$

$$+\frac{1}{2}\sum_{\kappa,\rho\in A}\sum_{\substack{\gamma,\nu\in AB\\ (\gamma\notin A)\vee(\nu\notin A)}}\left(D_{\kappa\gamma}D_{\rho\nu} - P^\alpha_{\kappa\nu}P^\alpha_{\rho\gamma} - P^\beta_{\kappa\nu}P^\beta_{\rho\gamma}\right) \times \left[[\gamma\nu|\kappa\rho] - \sum_{\tau,\eta\in A} A^A_{\gamma\tau}A^A_{\nu\eta}[\tau\eta|\kappa\rho]\right]$$

$$+\frac{1}{2}\sum_{\kappa,\rho\in B}\sum_{\substack{\gamma,\nu\in AB\\ (\gamma\notin B)\vee(\nu\notin B)}}\left(D_{\kappa\gamma}D_{\rho\nu} - P^\alpha_{\kappa\nu}P^\alpha_{\rho\gamma} - P^\beta_{\kappa\nu}P^\beta_{\rho\gamma}\right) \times \left[[\gamma\nu|\kappa\rho] - \sum_{\tau,\eta\in B} A^B_{\gamma\tau}A^B_{\nu\eta}[\tau\eta|\kappa\rho]\right],$$

that describe the difference between two-center integrals of some operators of intraatomic nature and their one-center approximations. These differences are due to the fact that functions like $\hat{h}^A\chi_\nu$ with $\nu \in A$ (and their two-electron analogues) can be approximated better when the basis of atom A is "extended" by adding the basis orbitals of atom B—hence the name "basis extension" terms. One can hardly attribute any true physical meaning to these terms; fortunately they diminish rather quickly as the basis set improves.

The first terms of (8.37) obviously describe true physical interactions: the nuclear-nuclear repulsion, the electron-nuclear attraction, and the interelectronic repulsion, respectively. As usual, when interelectronic interaction is considered, one can distinguish terms of Coulombic and exchange type. However, in the present case it may be meaningful to perform a little additional analysis [26,29], and distinguish also terms describing "pure" interatomic electrostatic (and respective exchange) effects and those of the *overlap* type. In this manner, we can identify the first terms in Eq. (8.37) as the sum of the electrostatic, overlap, and exchange contributions to the diatomic energy component E_{AB}:

$$E_{AB}^{el.stat.} + E_{AB}^{exch.} + E_{AB}^{overl.} = \frac{Z_A Z_B}{R_{AB}} \tag{8.55}$$

$$-\sum_{\tau\in AB}\left(\sum_{\mu\in A} B^{AB}_{\mu\tau} < \tau|\frac{Z_B}{r_B}|\mu > + \sum_{\mu\in B} B^{AB}_{\mu\tau} < \tau|\frac{Z_A}{r_A}|\mu >\right)$$

$$+\sum_{\substack{\kappa\in A\\ \rho\in B}}\sum_{\tau,\eta\in AB}[\tau\eta|\kappa\rho]\left(B^{AB}_{\kappa\tau}B^{AB}_{\rho\eta}-C^{\alpha AB}_{\kappa\eta}C^{\alpha AB}_{\rho\tau}-C^{\beta AB}_{\kappa\eta}C^{\beta AB}_{\rho\tau}\right)$$

In this approach we first consider the electron densities projected on the individual atoms and calculate their electrostatic interactions (along with the nuclei), as well as the respective exchange terms. The remainder of this projective approximation of the original terms in Eq. (8.37) will be attributed to overlap; this nomenclature will be justified below.

The electron density $\varrho(\vec{r})$ is given by Eq. (3.16); by grouping the terms with subscript μ according to the individual atoms, one can write

$$\varrho(\vec{r})=\sum_{\mu,\nu}D_{\mu\nu}\chi^*_\nu(\vec{r})\chi_\mu(\vec{r})=\sum_A\sum_{\mu\in A}\sum_\nu D_{\mu\nu}\chi^*_\nu(\vec{r})\chi_\mu(\vec{r}) \tag{8.56}$$

In order to approximate $\varrho(\vec{r})$ as a sum of "atomic" charge densities $\varrho'_A(\vec{r})$, expressed by the basis orbitals of the given atom only,* we replace every function $\chi^*_\nu(\vec{r})$ by its projection onto the subspace of the basis functions of the respective atom A:

$$\chi^*_\nu(\vec{r})\approx\left[\hat{P}^A\chi_\nu(\vec{r})\right]^*=\left[\sum_{\tau,\rho\in A}\chi_\tau(\vec{r})S^{-1}_{(A)\tau\rho}\int\chi^*_\rho(\vec{r}')\chi_\nu(\vec{r}')dv'\right]^*$$

$$=\sum_{\tau,\rho\in A}S_{\nu\rho}S^{-1}_{(A)\rho\tau}\chi^*_\tau(\vec{r})=\sum_{\tau\in A}A^A_{\nu\tau}\chi^*_\tau(\vec{r}) \tag{8.57}$$

Here we used the formula (2.1) for the projection operator, but without applying the "bra-ket" formalism. The last equality has been obtained by using the definition (8.38). Substituting this into Eq. (8.56) and using the definition (8.39), we get the approximation

$$\varrho(\vec{r})\approx\sum_A\varrho'_A(\vec{r}) \tag{8.58}$$

with

$$\varrho'_A(\vec{r})=\sum_{\mu,\tau\in A}B^A_{\mu\tau}\chi^*_\tau(\vec{r})\chi_\mu(\vec{r}) \tag{8.59}$$

It is easy to check by direct calculation that the "atomic" charge densities (8.59) integrate to Mulliken's gross atomic populations on the respective atoms. Therefore, their sum represents an approximation to the true

*The prime in $\varrho'_A(\vec{r})$ is used to distinguish this approximate quantity from the exact component $\varrho_A(\vec{r})$ considered in Eq. (3.28); the latter is not expressed solely by the orbitals of the given atom.

$\varrho(\vec{r})$ which conserves the total electronic charge of the molecule—a property extremely important from the point of view of the overall electrostatic balance. One can also see that the "atomic" charge density $\varrho'_A(\vec{r})$ defined here enters also the electron-nuclear attraction and the Coulombic part of the electron-electron repulsion terms of the *one-center* CECA energy components Eq. (8.36) derived essentially in a similar projective manner.

We obtain the electrostatic term of the interatomic energy component by determining the electrostatic interaction of the two nuclei and that of the "atomic" charge densities $\varrho'_A(\vec{r})$, and $\varrho'_B(\vec{r})$ with each other and with the nuclear charges Z_B and Z_A, respectively:

$$E_{AB}^{el.stat.} = \frac{Z_A Z_B}{R_{AB}} - \sum_{\mu,\tau\in A} B^A_{\mu\tau}\langle\tau|\frac{Z_B}{r_B}|\mu\rangle - \sum_{\mu,\tau\in B} B^B_{\mu\tau}\langle\tau|\frac{Z_A}{r_A}|\mu\rangle + \sum_{\tau,\kappa\in A}\sum_{\eta,\rho\in B} B^A_{\kappa\tau}B^B_{\rho\eta}[\tau\eta|\kappa\rho] \tag{8.60}$$

If the atoms A and B are far enough apart—they are divided by at least two or three bonds—then the electrostatic energy contribution $E_{AB}^{el.stat.}$ is usually rather close to the value one may obtain in the *point charge approximation*, by attributing to each atom the effective charge $Z_A - Q_A$, where Q_A is Mulliken's gross atomic population on that atom:

$$E_{AB}^{el.st.point} = \frac{(Z_A - Q_A)(Z_B - Q_B)}{R_{AB}} \tag{8.61}$$

The difference between the value of $E_{AB}^{el.st.point}$ from $E_{AB}^{el.stat.}$ describes the effect of the *deviation from the point charge*, so one may write

$$E_{AB}^{el.stat.} = E_{AB}^{el.st.point} + E_{AB}^{dev.point} \tag{8.62}$$

The term $E_{AB}^{dev.point}$ was called "penetration" in [26], by using a word from early quantum chemistry. It is usually pretty large for a pair of chemically bonded atoms, but—as already noted—its absolute values quickly decrease as the distance between atoms becomes larger. Comparing the terms of expression (8.60) with the respective ones in (8.55), one can see that they differ from each other by restricting some summation indices from the diatomic subspace AB to that of either atom A or atom B, and by replacing the elements of the diatomic "projected density matrix" $\mathbf{B}^{AB}$ with those of an atomic one $\mathbf{B}^A$ or $\mathbf{B}^B$. We attribute these differences to the overlap effects—see below. One can separate out the similar overlap effects in the exchange terms, too, by performing the same restriction on summation limits as discussed for the

Table 8.6
Diatomic energy components (in a.u.) of the ethane molecule computed by using 6-31G** basis set

Component	C-C	C-H	H-H geminal	H-H vicinal	C-H vicinal
E_{AB}	-0.5991	-0.6045	0.0157	0.0036 (2×) -0.0007 (1×)	0.0097
Electrostatic	-0.1407	-0.1500	0.0010	0.0020 (2×) 0.0015 (1×)	-0.0085
Point ch.	0.0388	-0.0182	0.0038	0.0026 (2×) 0.0021 (1×)	-0.0091
Dev. p. ch.	-0.1795	-0.1318	-0.0027	-0.0006 (2×) -0.0007 (1×)	0.0006
Exchange	-0.3353	-0.3305	0.0016	0.0000 (2×) -0.0004 (1×)	0.0025
Overlap	-0.0957	-0.0878	0.0149	0.0014 (2×) -0.0007 (1×)	0.0166
Basis ext.	-0.0274	-0.0361	-0.0018	0.0002 (2×) -0.0010 (1×)	-0.0010

*[29] with permission from Springer.

electrostatic terms, and analogously replacing the matrices $\mathbf{C}^{\sigma AB}$ with their atomic counterpart $\mathbf{C}^{\sigma A}$ or $\mathbf{C}^{\sigma B}$, similar to the case of matrices $\mathbf{B}^{X}$ in the electrostatic terms. Then we get the "pure" diatomic exchange energy term as

$$E_{AB}^{exch.} = -\sum_{\kappa,\tau\in A}\sum_{\rho,\eta\in B}\left(C_{\kappa\eta}^{\alpha B}C_{\rho\tau}^{\alpha A}+C_{\kappa\eta}^{\beta B}C_{\rho\tau}^{\beta A}\right)[\tau\eta|\kappa\rho] \tag{8.63}$$

This can indeed be considered the adequate expression for the diatomic exchange energy component: if one substitutes, similar to that discussed in Section 6.3, the asymptotic expansion $[\tau\eta|\kappa\rho] \approx S_{\tau\kappa}S_{\eta\rho}/R_{AB}$, then a trivial derivation utilizing the fact that $\sum_{\nu\in A}A_{\mu\nu}^{A}S_{\nu\rho} = S_{\mu\rho}$, if $\rho\in A$, shows that $E_{AB}^{exch.}$ becomes proportional to the bond order B_{AB}, as it should.

Extracting the right-hand sides of (8.60) and of (8.63) from that of (8.55), we get for the overlap terms the expression

$$E_{AB}^{overl.} = -\sum_{\tau\in AB}\left(\sum_{\mu\in A}B_{\mu\tau}^{AB}\langle\tau|\frac{Z_B}{r_B}|\mu\rangle+\sum_{\mu\in B}B_{\mu\tau}^{AB}\langle\tau|\frac{Z_A}{r_A}|\mu\rangle\right) + \sum_{\mu,\tau\in A}B_{\mu\tau}^{A}\langle\tau|\frac{Z_B}{r_B}|\mu\rangle+\sum_{\mu,\tau\in B}B_{\mu\tau}^{B}\langle\tau|\frac{Z_A}{r_A}|\mu\rangle \tag{8.64}$$

$$+\sum_{\substack{\kappa\in A\\ \rho\in B}}\sum_{\tau,\eta\in AB}[\tau\eta|\kappa\rho]\left(B^{AB}_{\kappa\tau}B^{AB}_{\rho\eta}-C^{\alpha AB}_{\kappa\eta}C^{\alpha AB}_{\rho\tau}-C^{\beta AB}_{\kappa\eta}C^{\beta AB}_{\rho\tau}\right)$$

$$-\sum_{\kappa,\tau\in A}\sum_{\eta,\rho\in B}[\tau\eta|\kappa\rho]\left(B^{A}_{\kappa\tau}B^{B}_{\rho\eta}-C^{\alpha B}_{\kappa\eta}C^{\alpha A}_{\rho\tau}-C^{\beta B}_{\kappa\eta}C^{\beta A}_{\rho\tau}\right)$$

It is easy to see that all terms of (8.64) are indeed due to the interatomic overlap: they either contain a diatomic "differential overlap" like $\chi^*_\mu(\vec{r})\chi_\nu(\vec{r})$; $\mu\in A,\ \nu\in B$ in the integral, or the interatomic elements of the inverse overlap matrix $\mathbf{S}^{-1}_{(AB)}$, or at least the differences like $S^{-1}_{(AB)\mu\nu}-S^{-1}_{(A)\mu\nu}$; $\mu,\nu\in A$ between the intraatomic elements of the diatomic and intraatomic inverse overlap matrices, which are also due to the non-zero interatomic block of the overlap matrix.

Tables 8.6 to 8.9 contain some decomposition of diatomic energy components obtained by using the well-balanced 6-31G** basis set [29]. It can be seen that the bonding contributions are dominated by the exchange components accounting for about half of the diatomic energy components; this stresses once again the close connection of exchange and bonding. The other three components (electrostatic, overlap, and basis extension ones) also are

Table 8.7
Diatomic energy components (in a.u.) of the ethylene molecule computed by using 6-31G** basis set

Component	C-C	C-H	H-H geminal	H-H vicinal	C-H vicinal
E_{AB}	-1.0915	-0.6108	0.0176	0.0045 0.0003	0.0124
Electrostatic	-0.2786	-0.1504	0.0018	0.0028 0.0019	-0.0078
Point ch.	0.0260	-0.0159	0.0047	0.0035 0.0028	-0.0082
Dev. p. ch.	-0.3046	-0.1345	-0.0028	-0.0007 -0.0009	0.0004
Exchange	-0.6237	-0.3321	0.0016	-0.0001 -0.0004	0.0030
Overlap	-0.1414	-0.0904	0.0162	0.0022 -0.0006	0.0206
Basis ext.	-0.0478	-0.0380	-0.0020	-0.0004 -0.0005	-0.0033

*[29] with permission from Springer.

negative and roughly comparable to each other; the electrostatic term being usually the largest. The point charge contribution to the electrostatic component is usually very small for the chemically bonded atoms, and the resulting value is determined by the actual distribution of the charges belonging to the interacting orbitals. In fact, the bonds are formed by atomic hybrid orbitals oriented toward each other; it is obvious that their electrostatic effect cannot be approximated by point charges placed at the nuclei. For non-bonded geminal, and vicinal atoms the overlap repulsion is the dominating effect, while for atoms which are more than one bond apart, the interaction often practically reduces to that of the respective Mulliken's gross atomic charges placed on the nuclei.

Table 8.8
Diatomic energy components (in a.u.) of the acetylene molecule computed by using 6-31G** basis set

Component	C-C	C-H	H-H vicinal	C-H vicinal
E_{AB}	-1.7607	-0.5641	0.0059	-0.0243
Electrostatic	-0.5343	-0.1346	0.0072	-0.0146
Point ch.	0.0243	-0.0273	0.0087	-0.0129
Dev. p. ch.	-0.5587	-0.1073	-0.0016	-0.0018
Exchange	-0.9846	-0.3043	-0.0004	-0.0057
Overlap	-0.1886	-0.0848	-0.0005	0.0039
Basis ext.	-0.0532	-0.0405	-0.0004	-0.0079

*[29] with permissionfrom Springer.

If the basis set improves, both the basis extension terms and (albeit more slowly) the overlap ones are reduced [29]. While the first seems natural, the second requires some discussion.

It is known that interatomic overlap effects are very important for understanding chemical bonding, in particular when qualitative discussions in terms of limited (minimal) atomic basis sets are concerned. The diatomic overlap terms $E_{AB}^{overl.}$ introduced here seem to reflect well these effects if adequate basis sets (e.g., 6-31G**) are used. At the same time, it is to be realized that the diatomic electrostatic term (8.60) is obtained by introducing an approximation of the electron density by the sum of "atomic densities" ϱ_A', and the errors of this approximation give rise to some overlap terms. Analogous projections enter, in fact, the exchange terms, so they also contribute to

Table 8.9
Some diatomic energy components (in a.u.) of the diborane molecule computed by using 6-31G** basis set

Component	B–$H_{br.}$	B–$H^a_{term.}$	B–B	$H^a_{br.}$–$H^a_{br.}$
E_{AB}	-0.2790	-0.5173	-0.2266	0.0253
Electrostatic	-0.0571	-0.1344	-0.0307	-0.0002
Point ch.	-0.0015	-0.0029	0.0050	0.0002
Dev. p. ch.	-0.0556	-0.1314	-0.0357	-0.0005
Exchange	-0.1425	-0.2978	-0.1296	-0.0034
Overlap	-0.0577	-0.0665	-0.0412	0.0240
Basis ext.	-0.0218	-0.0186	-0.0252	0.0050

[a] $H_{br.}$ and $H_{term.}$ denote hydrogen atoms in the bridge and terminal positions, respectively.
* [29] with permission from Springer.

$E^{overl.}_{AB}$. Obviously, the approximations involved become better and better as the basis set improves, so the terms $E^{overl.}_{AB}$ should diminish with improving basis sets. As they do reflect properly the role of interatomic overlap for small basis sets, giving negative contributions for bonded atoms and positive for situations with closed shell repulsions, they may be considered intermediate in their character between the terms describing "true" physical interactions and those that merely describe the incompleteness of the atomic basis sets. In fact, the meaning of "interatomic overlap" is lost if the basis on (at least) one atom becomes complete.

8.5 Improved CECA method

As noted above, the CECA method provides us an (exact or approximate) energy partition scheme that permits one to very clearly identify the chemical bonds corresponding to large negative diatomic energy components, the situations with overlap repulsion (small positive energy components), as well as other peculiarities of the molecular electronic structure—e.g., intramolecular hydrogen bonds or hydrogen-bond-like CH. . . O interactions [73,114]. It has a drawback, however, that the two-center bonding energy components are too large in absolute values, which one may feel uncomfortable. In fact, one obtains for the C–H and C–C *diatomic energy components* in the ethane molecule (using 6-31G** basis) the absolute values of about 375 kcal/mol, which is difficult to compare with the *bond dissociation energies* near 100 kcal/mol or even less. This uncomfortable feeling may remain even if one understands that these quantities have conceptually different physical meaning: the energy component is a static local quantity pertinent to only a given pair of atoms at the given geometry and wave function, while the dissociation energy is the global energy change taking place upon decomposing the molecule into two fragments. The overall energetics is basically in order, of course: the too negative bonding energy components are compensated by the one-center energy components that are "not negative enough"—they show much larger "promotion energies" than could be expected on the basis of the considerations discussed in Section 8.2.

Thus from the intuitive point of view one has too negative diatomic energy components and insufficiently negative atomic ones. Obviously, these are closely interrelated, and should be treated together if one wishes to get energy quantities that have numerical values that are more familiar to chemists. Of course, there is an infinite number of different decompositions of a single number (the total molecular energy) into the sum of different components (the one- and two-center contributions); our aim is to find one that permits to put the numbers "on the chemical scale" by conserving the good properties of the CECA scheme. For that reason we shall find the physical reasons behind the "non-chemical behavior" leading to too negative binding energy components and too large atomic promotion energies. To do that, we shall analyze the phenomenon in the simplest example of the H_2 molecule—first at the simplest semiempirical level and then at the *ab initio* level.*

The H_2 example

In the semiempirical MNDO ("modified neglect of diatomic overlap") method [116] the basis set is considered orthonormalized (which can be interpreted as if it is obtained from a tacit Löwdin-orthogonalization of the atomic minimal basis set) and one uses only one- and two-center integrals. By the proper semiempirical parameterization of the latter, a quite useful scheme emerged that can be used to study different molecular properties with a rather limited computational effort. As it contains only one- and two-center integrals, the MNDO energy can be trivially presented as a sum of one- and two-center energy components. These energy components have been used to

*Part of this section is reprinted from my paper [117] with permission from Royal Society of Chemistry.

discuss different details of electronic structure of molecules, among them to predict the primary cleavages in electron impact mass-spectrometry [75]. At the same time, the MNDO energy components have numerical values comparable to the CECA ones, i.e., also exhibit binding components that are too negative and one-center ones that are not negative enough.

Using MNDO, one obtains for the binding energy of the H_2 molecule at the experimental interatomic distance of 0.74 Å the quite reasonable value of 4.396 eV = 101.4 kcal/mol. However, the diatomic energy component is as large as -10.82 eV = -249.5 kcal/mol. This is compensated by the one-center energy components: each of them is higher than the MNDO energy of a free hydrogen atom by 3.212 eV = 74.07 kcal/mol. In accord with the fact that MNDO uses a minimal basis, the one-center energy components of the H_2 molecule *do not depend* on the interatomic distance. They consist of the expectation value of the one-electron atomic Hamiltonian that is equal to the MNDO energy of the free hydrogen atom, and an electron-electron repulsion term which is equal to the value 3.212 eV, mentioned above. The source of this number is obviously the following. Each hydrogen atom of the H_2 molecule contains one electron *on average*, so the one-electron energy component is equal to the MNDO atomic energy of the free hydrogen, but the use of doubly filled orbitals means that the wave function contains both covalent and ionic terms, and the latter give rise to some intraatomic electron repulsion which is absent in a free H atom. These ionic terms are responsible for the bad dissociation behavior of the RHF function at large distances but, of course, they are present at every distance, and we now identify them as the source of the imbalance in the one- and two-center energy components. In fact, one may conclude that these ionic energy components arise as *consequences of the bond formation,* therefore they should be assigned to the bond, rather than to the individual atoms: at the MNDO level there is no room for any true promotion of the hydrogen atom, so the hydrogen atom in the H_2 molecule differs from the free atom only by the fact of being bonded. The bonding does not justify the *one-center* energy to differ from that of the free atom.

We may apply this conclusion also to the minimal basis *ab initio* calculations of the H_2 molecule. Each H-atom has an average (gross) electron population equal to one, exactly as in the free atom. If a minimal basis is used, there is no room for any promotion in the hydrogen atom, so the intraatomic energy of that electron should not change with the bond formation and should not depend on the distance. If an energy decomposition scheme as applied to H_2 treated at the minimal basis level gives a one-center energy component differing from the free-atom energy calculated with the same basis set, then—similar to the MNDO case—this deviation should be reassigned to the interatomic bond, and not attributed to the individual atom.

We have started the discussion by considering the H_2 molecule at the MNDO level because the analysis in the *ab initio* case is much more involved, even if one uses a minimal basis with only one basis orbital on each hydrogen atom.

The one-electron Hamiltonian for the H_2 molecule can be written as

$$\hat{h} \equiv \hat{h}^A + \hat{U}^B \equiv \hat{h}^B + \hat{U}^A \tag{8.65}$$

Here $\hat{h}^A$ is the *intraatomic* Hamiltonian of atom A and $\hat{U}^B$ is the potential energy caused by nucleus B. Operators $\hat{h}^B$ and $\hat{U}^A$ are defined analogously. Assuming that the basis orbitals χ_1 and χ_2 are centered on atoms A and B, respectively, the expectation value of the one-electron Hamiltonian $\hat{h}$ can be presented in the form symmetric in the two atoms as

$$\begin{aligned}\langle \hat{h} \rangle = \sum_{\mu,\nu=1}^{2} D_{\mu\nu} h_{\nu\mu} &= D_{11}(h^A_{11} + U^B_{11}) + D_{12}(h^A_{21} + U^B_{21}) \\ &+ D_{21}(h^B_{12} + U^A_{12}) + D_{22}(h^B_{22} + U^A_{22}) ,\end{aligned} \tag{8.66}$$

where $D_{\mu\nu}$ are the elements of the usual density matrix. Thus the term containing $\hat{h}^A$ is $D_{11}h^A_{11} + D_{12}h^A_{21}$. In CECA the integral in the second term has been decomposed as

$$h^A_{21} = \langle \chi_2 | \hat{h}^A | \chi_1 \rangle = \langle \chi_2 | \hat{P}_A \hat{h}^A | \chi_1 \rangle + \langle \chi_2 | (1 - \hat{P}_A) \hat{h}^A | \chi_1 \rangle . \tag{8.67}$$

In the given case $\hat{P}_A = |\chi_1\rangle\langle\chi_1|$, thus we get the term containing $\hat{h}^A$ as

$$(D_{11} + D_{12}S_{21})h^A_{11} + \langle \chi_2 | (1 - \hat{P}^A) \hat{h}^A | \chi_1 \rangle = h^A_{11} + \langle \chi_2 | (1 - \hat{P}^A) \hat{h}^A | \chi_1 \rangle , \tag{8.68}$$

because $D_{11} + D_{12}S_{21}$ is Mulliken's *gross* atomic population on atom A and that is equal to one for the H_2 molecule. The term $\langle \chi_2 | (1 - \hat{P}_A) \hat{h}^A | \chi_1 \rangle$ would vanish if χ_1 were an exact eigenvector* of $\hat{h}^A$ or $\hat{P}_A$ corresponded to a complete basis. In CECA that term was assigned to the *diatomic* energy component—it contributes to a term of Eq. (8.37) in square braces—so for

*The exact eigenfunction of the hydrogenic $\hat{h}^A$ is, of course, known to be the 1*s* Slater orbital with exponent equal to 1, but it is not usually present in the basis sets used in quantum chemical calculations. (The standard STO-NG basis sets contain the approximation of the 1*s* Slater orbital with an exponent equal to 1.2, and not 1, in order to account for the "orbital shrinking" taking place during the bond formation.)

the monoatomic one-electron energy component the energy of the free hydrogen atom is recovered, as is required.

Now, we shall consider the one-center two-electron energy components, which in the H_2 case should be completely reassigned to the bond, instead of the individual atoms. In the closed shell case one has from the definitions (8.39) $\mathbf{C}^{\sigma A} = \frac{1}{2}\mathbf{B}$, so the two-electron part of the CECA one-center energy is obtained from Eq. (8.36) as

$$E_{(2)}^{A} = \frac{1}{2} \sum_{\mu,\nu,\rho,\tau \in A} (B_{\mu\nu}^{A} B_{\rho\tau}^{A} - \frac{1}{2} B_{\rho\nu}^{A} B_{\mu\tau}^{A}) [\chi_\nu \chi_\tau | \chi_\mu \chi_\rho] \,. \tag{8.69}$$

As there is only one basis orbital on each atom, a simple derivation shows that the same quantity can be expressed in the H_2 case simply as the integral

$$E_{(2)}^{A} = [\psi^A \psi^A | \varphi^A \varphi^A] \,, \tag{8.70}$$

where φ^A is the intraatomic part of the MO φ of the H_2 problem, and ψ^A is defined as the atomic projection of φ.*

One has to extract the quantity $E_{(2)}^{A}$ from the CECA atomic energy component of each atom and add it to the diatomic energy component. Then the requirement of recovering the free-atom energy for each H atom in the H_2 problem treated with a minimal basis will be fulfilled and the total energy —the sum of atomic and diatomic energy components—remains unchanged. The form (8.70) of the ionic energy contribution will be utilized in the following analysis.

The general closed shell case

In the H_2 case there is only one MO, so the identification of the ionic two-electron energy terms which are to be reassigned to the bond does not represent a problem. In many-atomic molecules there are several MOs. Some of them are doubly filled core or lone-pair orbitals; the ionic energy contributions corresponding to them, of course, represent intrinsic atomic characteristics and are by no means related to any bonding. There are also more or less localized bonding MOs for which corrections of the type considered for the H_2 case above should be quite appropriate—and there may be orbitals representing cases intermediate between these extremes (e.g., describe polarized bonds). In addition, there is the unitary freedom in selecting the occupied orbitals. In such a situation one needs a scheme in which the different terms

*Actually one has $\varphi^A = c_1^1 \chi_1$, $\psi^A = \hat{P}_A \varphi = (c_1^1 + S_{12} c_2^1)\chi_1$, where the c_μ^1-s are the orbital coefficients. In this minimal basis H_2 case $c_1^1 = c_2^1 = 1/\sqrt{2(1+S_{12})}$.

are not hand-picked arbitrarily but are determined automatically in a possibly optimal manner.

The theoretical tool we have found appropriate to treat all the complex situations which may occur in larger molecules is the use of the effective atomic orbitals and the connected with them molecular orbitals that were discussed in detail in Chapter 5. That means that one has to treat the ionic energy terms by determining for each atom a different set of occupied (localized) orbitals: that one which may be considered to be *privileged* from the point of view of the *given atom.* Thus, for each atom we perform the unitary transformation of the canonical orbitals $\varphi_j^c(\vec{r})$

$$\varphi_i^l(\vec{r}) = \sum_{j=1}^{occ.} U_{ji}\varphi_j^c(\vec{r}) \quad , \tag{8.71}$$

making the Magnasco–Perico localization criterion stationary, and diagonalizing the matrix $\mathbf{Q}$ with the elements

$$Q_{ij} = \sum_{\mu,\nu\in A} c_\mu^{i*} S_{\mu\nu} c_\nu^j \quad . \tag{8.72}$$

As discussed in Section 5.1, the privileged character of the orbitals φ_i^l is due to the fact that this is the only orthonormalized set of localized orbitals, for which the intraatomic parts of the orbitals also form an orthogonal atomic basis; the latter also diagonalize the *intraatomic* part of the density matrix. As noted in Section 5.1, for non-hypervalent atoms the number of localized orbitals with eigenvalues M_i significantly differing from zero always coincides with the number of classical minimal basis orbitals of that atom. Therefore, by this set of orbitals one can clearly distinguish the doubly filled orbitals, the bonding orbitals, and the practically empty basis orbitals of each atom. In this procedure, one can determine $min(m_A, n_{occ})$ orbitals for the given atom, where m_A is the number of basis functions on atom A and n_{occ} is the number of occupied orbitals in the molecule. (The "unnecessary" solutions, if any, correspond to zero eigenvalues M_i.) Of course, for determining these localized orbitals one may use any of the alternative algorithms discussed in Section 5.1 and in Ref. 40.

Similar to the case discussed for the H_2 molecule, for every MO φ_i^l corresponding to atom A we may define its intraatomic part φ_i^A

$$\varphi_i^A = \sum_{\mu\in A} c_\mu^{l,i} \chi_\mu \quad , \tag{8.73}$$

and its projection on the subspace of the basis orbitals centered on atom A

$$\psi_i^A = \sum_{\mu \in A} d_\mu^i \chi_\mu \quad , \tag{8.74}$$

where the coefficients d_μ^i can be expressed through the adjoint of matrix **A** defined in Eq. (8.38) as

$$d_\mu^i = \sum_\rho (\mathbf{A}^\dagger)_{\mu\rho} c_\rho^{l,i} \quad . \tag{8.75}$$

Every localized MO gives an "ionic" contribution to the one-center energy of atom A, which can be expressed, in terms of these notations, as

$$E_{(2)}^{Ai} = [\psi_i^A \psi_i^A | \varphi_i^A \varphi_i^A] , \tag{8.76}$$

in full analogy with Eq. (8.70) we had in the H_2 case. However, for the core or lone-pair orbitals these ionic terms are inherently of atomic nature, while for the bonding orbitals they should be attributed to the bond(s). Instead of considering that question on a case-to-case basis, we introduce the following general treatment for solving that problem in a general fashion, permitting the intermediate cases to be also treated appropriately.

A doubly filled atomic orbital or an empty atomic orbital does not contribute to the *valence* V_A of the atom Eq. (6.57), while an orbital forming a non-polarized bond contributes one to that valence. Therefore, we multiply the integral in Eq. (8.76) by the contribution V_A^i of the corresponding effective atomic orbital χ_i' (normalized version of the intraatomic component φ_i^A) to the total valence, and extract the product $V_i^A[\psi_i^A \psi_i^A | \varphi_i^A \varphi_i^A]$ from the one-center energy—and add it to the two-center energy components, distributed according to the bond orders formed by the effective AO in question.

The problem of partial valences has been discussed in Section 6.12, so we only write here that the total valence of the atom can be presented as a sum of partial valences corresponding to the individual basis functions—in the given case of the effective atomic orbitals χ_i':

$$V_A = \sum_i V_A^i \quad , \tag{8.77}$$

where V_A^i is the contribution of the i-th effective atomic orbital to the total atomic valence. It is equal to

$$V_A^i = 2(\mathbf{D}'\mathbf{S}')_{ii} - \sum_{j \in A} (\mathbf{D}'\mathbf{S}')_{ij} (\mathbf{D}'\mathbf{S}')_{ji} \quad , \tag{8.78}$$

Table 8.10
Energy components of the methane ethane, ethylene, and acetylene molecules, calculated by using the 6-31G** basis set

Atom(s)	Energy component (kcal/mol)			
	Methane	Ethane	Ethylene	Acetylene
C	59.22	83.91	108.28	155.38
H	57.61	54.62	64.76	102.57
C–C	—	-159.13	-296.59	-481.02
C–H	-168.13	-167.20	-167.22	-150.97
H–H geminal	8.21	8.58	10.05	—
H–H vicinal	—	2.30 (2x) 1.54 (1x)	2.97 1.90	5.03
C–H vicinal	—	5.92	6.51	-16.80

*[117] with permission from Royal Society of Chemistry.

where $\mathbf{D}'$ and $\mathbf{S}'$ being the density and overlap matrices, respectively, transformed to the new basis orbitals χ_i' used on atom A. (The basis orbitals on the other atoms may be considered unchanged.)

As is discussed in Section 6.12, the partial valence is a sum of partial bond orders the given orbital forms with all the other atoms:

$$V_A^i = \sum_{\substack{B \\ B \neq A}} B_{AB}^i \quad ; \tag{8.79}$$

as a consequence of the idempotency relationship $(\mathbf{DS})^2 = 2\mathbf{DS}$ valid for the closed-shell determinant wave functions. The partial bond order B_{AB}^i corresponding to the effective AO χ_i' of atom A can be written as

$$B_{AB}^i = \sum_{\nu \in B} (\mathbf{D}'\mathbf{S}')_{i\nu} (\mathbf{D}'\mathbf{S}')_{\nu i} \tag{8.80}$$

As shown in [5] (c.f. also the discussion on page 39), the matrix product **DS** transforms once contravariantly and once covariantly, so the transformed product $\mathbf{D}'\mathbf{S}'$ can be expressed in terms of the original matrices as

$$\mathbf{D}'\mathbf{S}' = \mathbf{T}^{-1}\mathbf{DST} \quad , \tag{8.81}$$

where matrix **T** is the matrix describing the transformation from the original basis functions χ_μ to the new ones:

$$\chi_i' = \sum_{\mu} T_{\mu i} \chi_\mu \quad . \tag{8.82}$$

Table 8.11
Energy components of the benzene molecule calculated by using the 6-31G** basis set

Atom(s)	Energy component (kcal/mol)
C	162.62
H	73.69
C_1–C_2	-291.46
C_1–C_3	18.45
C_1–C_4	21.52
C_1–H_7	-157.23
C_1–H_8	4.11
C_1–H_9	-1.55
C_1–H_{10}	-1.22
H_7–H_8	2.92
H_7–H_9	1.87
H_7–H_{10}	1.37

*[117] with permission from Royal Society of Chemistry.

In the present case matrix **T** is a unit matrix, except for its block connecting the new and old basis orbitals on atom A, and the same holds for its inverse. Because the "effective" AOs are orthonormalized, one has for the *AA*-block of matrix **T** the equality

$$\mathbf{T}^{AA\dagger}\mathbf{S}^{AA}\mathbf{T}^{AA} = \mathbf{I}^{AA} \quad , \tag{8.83}$$

from which

$$(\mathbf{T}^{AA})^{-1} = \mathbf{T}^{AA\dagger}\mathbf{S}^{AA} \quad , \tag{8.84}$$

which permits the partial bond orders B^i_{AB} to be calculated in an easy fashion as

$$B^i_{AB} = \sum_{\nu\in B}(\mathbf{T}^\dagger\mathbf{SDS})_{i\nu}(\mathbf{DST})_{\nu i} \quad , \tag{8.85}$$

where it has been taken into account that $T_{\mu\nu} = T_{\nu\mu} = \delta_{\mu\nu}$ if $\nu \in B$; $B \neq A$. The partial valences V^i_A need not be calculated explicitly by using Eq. (8.78), but may be simply obtained as sums according to (8.79).

Thus the final one- and two-center energy components are given [117] by

$$E_A = E^{CECA}_A - \sum_{i\in A}[\psi^A_i\psi^A_i|\varphi^A_i\varphi^A_i]V^i_A \quad , \tag{8.86}$$

Table 8.12
Energy components of the diborane molecule calculated by using the 6-31G** basis set

Atom(s)	Energy component (kcal/mol)
B	240.37
H bridge	-0.54
H terminal	-9.51
B–H bridge	-96.79
B–H terminal	-127.38
B–B	-112.62
H bridge–H bridge	27.73
H bridge–H terminal	2.44
H terminal–H terminal	3.71
H terminal–H terminal′	0.97
H terminal–H terminal″	0.76

*[117] with permission from Royal Society of Chemistry.

and

$$E_{AB} = E_{AB}^{CECA} + \sum_{i \in A} [\psi_i^A \psi_i^A | \varphi_i^A \varphi_i^A] B_{AB}^i \quad , \tag{8.87}$$

where E_A^{CECA} and E_{AB}^{CECA} are given either by the approximate CECA expressions (8.36) and (8.37), or by the "exact" ones (8.47) and (8.48) corrected for the remaining three- and four-center effects absent in the original CECA.

In the actual calculations we have performed, these corrections were applied to the "exact" version of CECA. They were done by our freely available program [118] which is applicable at the closed-shell HF level* for basis sets containing s-, p-, and d-orbitals.

Tables 8.10 – 8.13 show the results of some illustrative calculations. One may see that the improved CECA scheme apparently fulfills all the expectations, and the results may be called very "chemical." The energy component corresponding to a C–C or C–H single bond is obtained about 150–170 kcal/mol (in absolute value), significantly less than the too large values (350–400 kcal/mol) of the CECA (and E1) schemes. The non-bonded interactions are not much influenced by the corrections introduced here and we get results similar to those in CECA. In the cases where some significant difference oc-

*The formalism of "ionic" corrections has not been generalized to open-shell systems, as yet.

curs, the present results seem much more reassuring: it was not easy to understand why one of the vicinal H...H interactions gave a small negative number in the ethane molecule [111]—here no such problem arises.

While the C–H energy components do not change too much from molecule to molecule, the one-center energy components—especially those of the carbon atoms—vary significantly in the hydrocarbons studied. (The differences ΔE_A with respect to the free ROHF atoms are always displayed.) In the previous studies practically no attention has been paid to the changes in the one-center energy components—they were either too far from or too close to the free atomic energies, to be meaningfully compared. The new scheme gives a new possibility in this respect, too. Besides the atomic promotions discussed in Section 8.2, the main factor seems to be the following: an atom "collects" electrons during the molecule formation if that is favorable energetically. This energetic effect manifests in the one-center energy contributions: as a tendency, the more negative atoms exhibit lower one-center energy components.

The connection between the one-center energies and the atomic charges appears even more prominent for systems also containing heteroatoms. Table 8.13 displays some results for ethanol, glycine, and N,N-dimethylformamide, $(CH3)_2N\text{-}CHO$, molecules: the one-center energy components and the bonding two-center components are presented. (Where there are more than two atoms/bonds of a given type, then only the limiting values are displayed in parentheses.) One can see that the numbers obtained do completely agree with the chemical expectations: the different atoms exhibit one-center components characteristic to their nature and chemical environment: electronegative atoms (oxygen, nitrogen) have negative one-center components, carbon and hydrogen have positive ones; the more positive the atom, the higher is its one-center energy. This leads to big differences between carbon and hydrogen atoms in the same molecule. Chemically similar atoms have, of course, similar one-center energies. Thus the two hydrogens in the CH_2 group of the ethanol molecule have identical one-center energies of 45.7 kcal/mol, while the CH_3 group has two equivalent hydrogens (62.2 kcal/mol) and one which differs somewhat (53.0 kcal/mol). Similar tendencies hold for the C–H two-center energy components, too: -155.5 kcal/mol in the CH_2 group and -170.1 (twice) and -166.0 kcal/mol in the CH_3 one, respectively. Of course, the hydrogen atom in the OH group is positive, which leads to a rather high one-center energy component. The interaction between the positive hydrogen with the negative oxygen leads to an enhanced O–H two-center energy component, as compared with the C–H ones.

Table 8.13
One-center and bonding two-center energy components of the ethanol, glycine, and N,N,dimethylformamide molecules calculated by using the 6-31G** basis set

Atom(s)	Energy component (kcal/mol)		
	Ethanol	Glycine	Dimethylformamide
C (CH_3)	84.11	—	—
C (CH_2)	228.72	155.58	—
C (COOH)	—	612.60	—
C (CHO)	—	—	507.15
O (COH)	-163.59	-246.8	—
O (C=O)	—	-212/39	-199.42
N	—	-28.93	-140.54
H (CH)	(45.7; 62.2)	61.5; 74.8	(50.7; 91.7)
H (OH)	209.22	225.00	—
H (NH)	—	149.1; 160.9	—
C–C	-183.69	-166.92	—
C–O (COH)	-145.70	-223.98	—
C–O (C=O)	—	-353.06	-339.54
N–C (NH_2CH_2)	—	-170.21	—
N–C (CH_3N)	—	—	-176.8; -180.5
N–C (NCO)	—	—	-280.00
C–H (CH_3, CH_2)	(-155.5; -170.1)	-165.0; -172.2	(-157.2; -163.2)
C–H (CHO)	—	—	-118.79
O–H	-260.44	-250.20	—
N–H	—	-233.0; -235.1	—

*[117] with permission from Royal Society of Chemistry.

8.6 Decomposition of the correlation energy

In this book the attention is mainly attributed to the calculations performed at the SCF (Hartree–Fock–Roothaan) level. This is because *most of the chemistry* can be understood at the Hartree–Fock level; if we are interested in the main qualitative aspects of the molecular structure (especially of ordinary organic compounds at their equilibrium conformations) and not in accurate values of small energy differences, then we need basically to analyze the results obtained in the HF calculations. This is the case, despite the fact that correlation energy is quite comparable with the binding energies representing interest from the chemical point of view. However, the correlation energy can be presented as a sum of components—the so-called *pair correlation energies*—that can be assigned to the individual HF molecular orbitals, which do not vary too much in similar conditions. However, if desired, one can decompose the correlation energy into atomic and diatomic contributions, based just on its presentation as a sum of pair correlation energies.

It may be noted here that for variational wave functions (e.g., CI, MC-SCF, CASSCF) one could proceed, in principle, quite analogously to the SCF case. The general form of the energy expectation value is

$$E = \sum_{\mu,\nu} D_{\mu\nu} h_{\nu\mu} + \sum_{\mu,\nu,\varrho,\tau} \Gamma_{\mu\varrho\nu\tau} [\chi_\nu \chi_\tau | \chi_\mu \chi_\varrho] + \sum_{A<B} \frac{Z_A Z_B}{R_{AB}} \tag{8.88}$$

where $\mathbf{D}$ and $\mathbf{\Gamma}$ are the actual spin-less first- and second-order density matrices. Then one can introduce one or two "atomic resolutions of identity" to the "kets" in the integrals, similarly to that made in Section 8.4.4 for the SCF case. However, when applying the most important perturbational and coupled-cluster methods, we do have a wave function, but the energy is not calculated as any expectation value. Nesbet's theorem used in the present section is applicable for both variational and non-variational methods.

To begin with, we should consider Nesbet's theorem [5,119]. Let us write the (exact or approximate) correlated wave function Ψ in the so-called correlation normalization:

$$\Psi = \Phi_0 + \phi \; , \tag{8.89}$$

where Φ_0 is the HF solution for the system considered and ϕ is the correlation correction. The correlation normalization means that one requires

$$\langle \Phi_0 | \Phi_0 \rangle = 1 \; ; \quad \langle \Phi_0 | \phi \rangle = 0 \; . \tag{8.90}$$

Substituting the expression (8.89) into time-independent Schrödinger equation $\hat{H}\Psi = E\Psi$, we can write

$$\hat{H}(\Phi_0 + \phi) \cong (E_0 + \Delta E)(\Phi_0 + \phi) \; . \tag{8.91}$$

Here E_0 is the Hartree–Fock energy, ΔE the correlation energy, and the symbol "$\cong$" indicates that usually we are dealing only with an approximate solution of the Schrödinger equation. Multiplying Eq. (8.91) with Ψ_0^* (from left) and integrating, we get—taking into account that $\langle \Phi_0 | \hat{H} | \Phi_0 \rangle = E_0$:

$$\Delta E \cong \langle \Phi_0 | \hat{H} | \phi \rangle \; . \tag{8.92}$$

Now, using the orthonormalized sets of occupied and virtual HF spin-orbitals $\psi_i,\ \psi_j\ldots$ and $\psi_u,\ \psi_v\ldots$, respectively, the correlation component ϕ of the wave function can be expressed as a linear combination of determinants that are singly, doubly, etc., excited with respect to the HF "ground state" Φ_0:

$$\phi = \sum_{i}^{occ.}\sum_{u}^{virt.} c_i^u \Phi_1(\psi_i \to \psi_u) + \sum_{i<j}^{occ.}\sum_{u<v}^{virt.} c_{ij}^{uv}\Phi_2(\psi_i \to \psi_u; \psi_j \to \psi_v)\ldots \quad (8.93)$$

Substituting this expansion into the expression (8.92), the singly excited determinants do not interact with the HF ground state Φ_0 due to the known Brillouin theorem, while those which contain triply or more excited determinants do not interact due to the Slater rules [5]. Thus only doubly excited determinants contribute to the correlation energy. Each doubly excited determinant gives one two-electron integral and its exchange counterpart. Thus we get the expanded form of Nesbet's theorem:

$$\Delta E = \sum_{i<j}^{occ.}\sum_{u<v}^{virt.} c_{ij}^{uv}\left([\psi_i\psi_j|\psi_u\psi_v]' - [\psi_i\psi_j|\psi_v\psi_u]'\right) . \quad (8.94)$$

Here the primes indicate that the integrals contain summations over the spin variables.

One can group the terms of this expression in a natural manner and write it as a sum of pair-correlation energies ε_{ij} corresponding to the pair of occupied spin-orbitals* ψ_i and ψ_j

$$\Delta E = \sum_{i<j}^{occ.} \varepsilon_{ij} , \quad (8.95)$$

where

$$\varepsilon_{ij} = \sum_{u<v}^{virt.} c_{ij}^{uv}\left([\psi_i\psi_j|\psi_u\psi_v]' - [\psi_i\psi_j|\psi_v\psi_u]'\right) . \quad (8.96)$$

Now, it is a natural choice to introduce two "atomic resolutions of identities" into the "bras" of the two-electron-integrals in this expression.† One then obtains

$$\varepsilon_{ij} = \sum_{A,B}\sum_{u<v}^{virt.} c_{ij}^{uv}\left([\hat{\rho}_A\psi_i\hat{\rho}_B\psi_j|\psi_u\psi_v]' - [\hat{\rho}_A\hat{\rho}_B\psi_i\psi_j|\psi_v\psi_u]'\right) . \quad (8.97)$$

*If Ψ_0 is an RHF or ROHF wave function, one could also combine the terms corresponding to the same spatial orbitals and different spins. This interesting aspect is, however, out of our present scope.

†In the previous sections we usually applied "atomic resolutions of identities" in the "kets" of different integrals. Here we could use that approach, too, by interchanging the "bra" and "ket" sides in Eq. (8.92), which would not change anything. The form given in Eq. (8.92) is, however, the widely used one, so we decided to stick to it.

Thus one obtains

$$\Delta E = \Delta E_{AB} \,, \tag{8.98}$$

where

$$\Delta E_{AB} = \sum_{i<j}^{occ.} \sum_{u<v}^{virt.} c_{ij}^{uv} \left([\hat{\rho}_A \psi_i \hat{\rho}_B \psi_j | \psi_u \psi_v]' - [\hat{\rho}_A \psi_i \hat{\rho}_B \psi_j | \psi_v \psi_u]' \right) \,. \tag{8.99}$$

In the Hilbert-space analysis, the two-electron integrals in this expression can be expanded in terms of the integrals over the basis functions and of the orbital coefficients $C_{\mu p}^{\sigma}$ as

$$[(\hat{\rho}_A \psi_i)(\hat{\rho}_B \psi_j) | \psi_u \psi_v]' = \sum_{\mu \in A} \sum_{\nu \in B} \sum_{\varrho, \tau} C_{\mu i}^{\sigma_i} C_{\nu j}^{\sigma_j} C_{\varrho u}^{\sigma_i} C_{\tau v}^{\sigma_j} [\chi_\mu \chi_\nu | \chi_\varrho \chi_\tau] \,, \tag{8.100}$$

if the spins of the virtual orbitals satisfy $\sigma_u = \sigma_i$, $\sigma_v = \sigma_j$, and zero otherwise. (For the second integral in (8.99) one has to interchange u and v both in the expansion (8.100) and in the conditions for spins.)

In the case of second order UMP theory, that has already been considered in practice [100], the coefficient c_{ij}^{uv} is given by

$$c_{ij}^{uv} = \frac{[\psi_u \psi_v | \psi_i \psi_j]' - [\psi_v \psi_u | \psi_i \psi_j]'}{\varepsilon_u + \varepsilon_v - \varepsilon_i - \varepsilon_j} \,. \tag{8.101}$$

The decomposition of the correlation energy is somewhat troublesome because formula (8.100) means that—besides a standard integral transformation—one has to perform a partial integral transformation for every atom and pair of atoms. Thus the cost of such calculations sharply increases with the dimensions of the system.

It may be interesting to note that Ayala and Scuseria [120] have decomposed into one- and two-center terms the coefficients (8.101) and not the matrix element (integrals) in the Nesbet expression (8.92). This corresponds to a completely different philosophy than ours, because essentially the excitations are attributed to the different atoms or pairs of atoms, and the integral describing the interactions is not decomposed. However, owing to the analogy of the expressions (8.96) and (8.101) in the MP2 case they give *exactly* the same results. At the same time, these approaches may lead to different results for other theories. It is our opinion that the present approach is to be preferred in such a case, because it is consistent with the practice used in decomposing the SCF energy: not the wave function but the interaction integrals are partitioned.

Table 8.14
Correlation energy components for the ethane molecule calculated at the MP2 level by using different basis sets (Core electrons are included.)

Basis set	Energy components (a.u.)			
	C	H	C–C	C–H
6-31G	-0.0645	-0.0059	-0.0039	-0.0031
6-311G	-0.0833	-0.0061	-0.0045	-0.0038
6-31G*	-0.0869	-0.0060	-0.0161	-0.0073
6-311G*	-0.1048	-0.0060	-0.0179	-0.0085
6-31G**	-0.0876	-0.0084	-0.0162	-0.0112
6-311G**	-0.1006	-0.0087	-0.0183	-0.0128
6-311++G**	-0.1022	-0.0087	-0.0172	-0.0132
6-311G(2d,p)	-0.1028	-0.0089	-0.0205	-0.0137

*[100] permission from Royal Chemical Society.

Table 8.14 contains some numerical results performed for the ethane molecule by using different basis sets [100]. One can see that the correlation energy components along the chemical bonds ("left–right correlation") change slowly with the basis sets; the most important factor is the presence or absence of polarization functions. The intraatomic correlation components describing the "in-out" and "angular" correlations are significant for heavy atoms and increase quickly in absolute value as the basis set improves. But they are inherently atomic quantities, so we can conclude that the correlation energy components do not carry much chemical information (at least at the equilibrium molecular conformation).

9

Analysis in the three-dimensional space

As noted in the Introduction, there are two conceptually different approaches to define an atom *within* a molecule, leading to the "Hilbert space" analysis and the analysis in the three-dimensional (3D) physical space. Classical quantum chemistry was based on the LCAO concept which later developed into the use of (larger and larger) atom-centered basis sets as the standard tools of calculations. Therefore—started with the simple Hückel theory—the methods of Hilbert-space analysis were first developed historically. Hilbert space analysis also has the advantage that it usually does not require calculation of any new quantities, only some combination of those that are calculated anyway when computing the wave function. Accordingly, Hilbert space analysis got the main attention in the present book, too. However, as noted on page 14, one may expect adequate results only if some "well-balanced" basis sets are used. Applying methods of Hilbert space analysis, one may meet irrecoverable difficulties if the basis set contains diffuse functions lacking any true atomic character. For anions, in particular, the use of diffuse functions is mandatory, and their treatment by using Hilbert-space based indices is virtually impossible. (The same holds to a large extent also to intermolecular interactions, for which obtaining any quantitative results requires the use of diffuse functions.) From a conceptual point of view, it also represents a problem that the results of Hilbert space analysis are lacking any basis set limits. Last but not least, the methods of Hilbert space analysis are not applicable to the calculations using plane wave basis, which recently became widespread.*

These drawbacks are avoided if one uses 3D-analysis. In the 3D case one decomposes the physical space into atomic domains with sharp or "fuzzy" boundaries. Experience shows that in such a frame the different quantities converge well (or even very well) as the basis sets improve; there appears, however, the problem that the division of the 3D space into atomic domains contains an obvious element of arbitrariness. (This is in agreement with the

*There are attempts to circumvent this problem by performing projections of the orbitals obtained in the plane wave calculations on some predefined local basis and performing analyses of Hilbert-space type in terms of the latter.

fact that the concept of an atom *within* a molecule cannot be given a strict definition.)

The present chapter gives a short overview of how the different quantities discussed previously can be calculated in the framework of the 3D analysis. The task is much simplified by the fact that the formalism of atomic decomposition of identity discussed in Chapter 3 permits to transform the formulae of one formalism to those of another, performing essentially some substitutions. In fact, comparison of Eqs. (3.9) and (3.11) indicates that one should perform the replacements like

$$\sum_{\nu\in A} c_\nu \langle \ldots|\chi_\nu\rangle \longrightarrow \sum_{\nu} c_\nu \langle \ldots|w_A(\vec{r})\chi_\nu(\vec{r})\rangle \,, \tag{9.1}$$

and similarly for the adjoint quantities

$$\sum_{\mu\in A} c_\mu^* \langle \chi_\mu|\ldots\rangle \longrightarrow \sum_{\mu} c_\mu^* \langle \chi_\mu(\vec{r})w_A(\vec{r})|\ldots\rangle \,. \tag{9.2}$$

Of course, the analysis of different quantities in the 3D framework can also be performed directly, by introducing one or more times the decomposition of identity as

$$\sum_A w_A(\vec{r}) = 1\,, \quad \forall \vec{r}\,, \tag{9.3}$$

taking care that quantum mechanical operators should not act on the weight functions $w_A(\vec{r})$—c.f. the discussion in Section 3.2.

9.1 Different weight functions

As has been mentioned in Section 3.2, there are two conceptually different ways of defining atomic domains in the 3D space: they either can be disjunct or "fuzzy" ones. In the disjunct case every point of space is assigned to one, and only one, atomic domain Ω_A. This situation is described by Eq. (3.14):

$$w_A(\vec{r}) = \begin{cases} 1 \text{ if } \vec{r}\in\Omega_A\,, \\ 0 \text{ otherwise}\,; \end{cases} \tag{9.4}$$

This is characteristic for the AIM method of Bader [2], in which the atomic domains are determined by the actual topology of the electronic charge distribution: they are delimited by surfaces that represent sorts of "watersheds" that divide pieces of charge "attracted" to that or another center. These centers usually coincide with the nuclei, but sometimes there appear spurious "non-nuclear attractors," too, representing the greatest conceptual weakness of Bader's AIM theory.

From the practical point of view, the largest disadvantages of the AIM method are the large CPU demand of the detailed investigation of the topology of the electron density and the observation that the form of the atomic domains does not always agree with our physical intuition, which does not help the understanding of the results.

Another obvious drawback of a theory decomposing the space into disjunct atomic domains is that the electronic charge at any point $\vec{r}$ is attributed to one, and only one, of the atoms. Thus it cannot reflect *directly* the well-known fact that there is an accumulation of electronic charge density between those atoms that are chemically bonded. This means that there is a meaning to assign a part of the charge not to a given atom but to a *pair* of atoms. However, in Bader's scheme all the overlap populations Eq. (3.22) are strictly zero, because in this case $\hat{\rho}_B^\dagger \hat{\rho}_A = w_B(\vec{r}) w_A(\vec{r}) = 0$ as the point $\vec{r}$ cannot belong to two *disjunct* atomic domains simultaneously.* (One may disregard the points on the delimiting surfaces between the domains as they represent "sets of measure zero.") This explains that from a chemical point of view there is a meaning to introduce "fuzzy" atomic domains, which means that the regions assigned to the individual atoms have no sharp boundaries but exhibit a continuous transition from one to another.

The first definition of the "fuzzy" atoms was introduced by Hirshfeld [3]. His scheme (often called the "stockholder's method") is based on the following considerations. One at first defines the so-called "promolecule" that is imagined as a superposition of free atoms put in the positions which the different atoms have in the molecule. In each point of the space one computes an auxiliary quantity, the electron density of the promolecule, as the sum of the free-atom densities $\varrho_A^0(\vec{r})$ at the given point. The weight function of a given atom to be used in the 3D analysis is then identified by the ratio by which the given atom contributes to the promolecule density:

$$w_A(\vec{r}) = \frac{\varrho_A^0(\vec{r})}{\sum_B \varrho_B^0(\vec{r})} \tag{9.5}$$

It is easy to see that this definition satisfies the requirement (9.3).

*There are some tools in the AIM theory, too, permitting to discuss the accumulation of the charge in the bonding regions, in particular the investigation of the "bond critical points." Their discussion is, however, out of our scope.

Hirshfeld's scheme is still widely used. It has a refined version, too, called "iterative Hirshfeld" (or "iterative stockholder's") method [121]. It takes into account that in some cases it is difficult to determine whether the molecule should be considered as a system of bonded *neutral atoms*, or some atoms are basically in an ionized state (e.g., H^+F^- rather than HF). In the "iterative Hirshfeld" method each "free atom of the promolecule" is considered as a superposition of the neutral atom and an ion. In the "iterative Hirshfeld" scheme the density of the given atom A in the promolecule, used in Eq. (9.5) is calculated as

$$\varrho_A^0(\vec{r}) = x_A \cdot \varrho_A^{0(neutr.)}(\vec{r}) + (1 - x_A) \cdot \varrho_A^{0(ion)}(\vec{r}) \;, \tag{9.6}$$

and iterations* are performed until the net atomic charge $Z_A - Q_A$ becomes equal to the value $\pm(1 - x_A)$ used in defining the weight function. (The sign obviously depends on whether one has to consider the positive or the negative ion besides the neutral atom. Z_A is the nuclear charge and Q_A is the electron population discussed in the next section.)

A conceptually different type of weight functions is the use of some analytical formulae to determine the delimitation and continuous transition of the spatial domains. In the work [122] we have proposed to use for that purpose Becke's procedure [123], originally introduced for performing the numerical integrations in the DFT calculations. The only external parameters it uses are the standard radii of the atoms; one has also to select an integer parameter defining the stiffness of delimiting the atomic domains. Becke's weight functions not only satisfy the condition (9.3), but also possess the property, that for every atom its weight function is equal to one on its "own" nucleus and zero on the nuclei of other atoms:

$$w_A(\vec{R}_A) = 1 \;; \qquad w_A(\vec{R}_B) = 0 \;, \; (A \neq B) \;. \tag{9.7}$$

$\vec{R}_A, \vec{R}_B$ are the positions of the nuclei A and B, respectively. The procedure of determining Becke's weight function is given in Appendix A.2

In the case of using Becke functions, the problem of neutral *vs.* ionized states of atoms does also appear, because the algorithm uses the atomic radii entered as input data. For instance, in the paper [122] we had to apply for fluorine the average of the atomic and ionic radii to get reasonable results. A

*One starts with the values $x_A = 1$ corresponding to the neutral atoms, then calculates the atomic net charges $Z_A - Q_A$, from which one decides whether the positive or the negative ion is to be considered for the given atom, determines the next value of x_A from the relationship $1 - x_A = |Z_A - Q_A|$, and repeats the calculation with the values obtained in this manner until convergence is reached.

better solution of that problem is qualitatively similar to the iterative Hirshfeld method in the aspect that the actual densities *in the given molecule* are taken into account. In this scheme [124, 125] called "fuzzy-rho" or "Becke-rho," one utilizes the fact that the atomic radii enter the Becke's algorithm only through their ratios. For any pair of chemically bonded atoms one can find the minimum of density along the line connecting the nuclei, and determine the ratio of the atomic radii based on the position of this minimum. (For atoms that are not chemically bonded it is practically indifferent what radii are used.) Obviously, this approach shows some conceptual similarity to Bader's scheme. too.

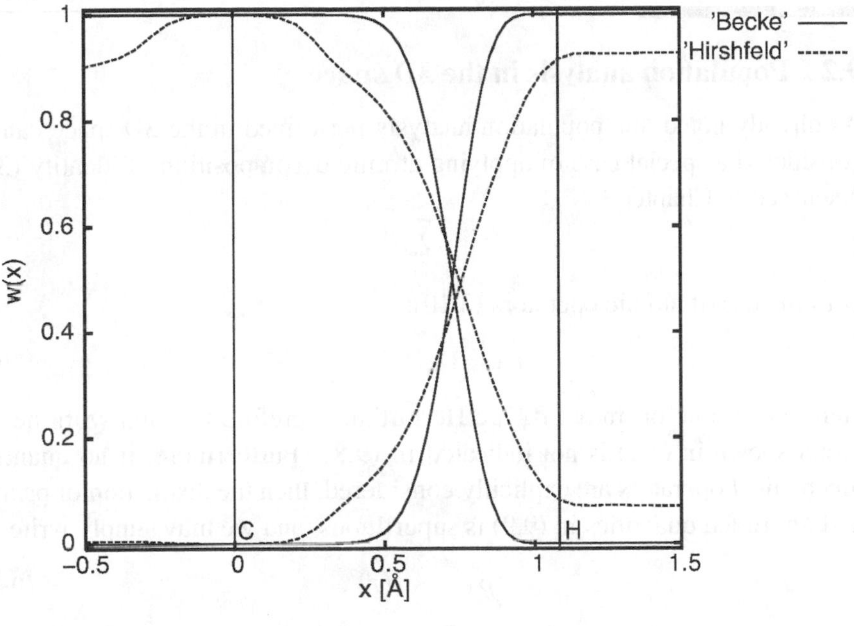

Figure 9.1

Becke and Hirshfeld weight functions along the line connecting the carbon and hydrogen atoms in the methane molecule. [100] permission from Royal Chemical Society.

Figure 9.1 compares the Becke and Hirshfeld weight functions along the straight line connecting the carbon and hydrogen atoms in the methane molecule. (The value of parameter $k = 3$, suggested by Becke, has been used for calculating the Becke function.) It can be seen that the point when the weight functions of carbon and hydrogen cross each other (have the values equal 1/2) are almost at the same point for the two schemes, dividing the interatomic distance in the ratio close to 2/3 to 1/3, but their qualitative behavior near the proton is completely different. The Hirshfeld curve for carbon exhibits a small but in no way negligible value at the proton, while the Becke one is exactly zero there. This means that a part of the electronic charge near the hydrogen is still attributed to the carbon in the Hirshfeld scheme. No such phenomenon occurs for carbon. This difference is obviously a consequence of the absence of core orbitals for hydrogen.

9.2 Population analysis in the 3D space

As already noted, the population analysis performed in the 3D space can be considered a special case of applying atomic decomposition of identity (3.1) discussed in Chapter 3

$$\hat{I} = \sum_A \hat{\rho}_A \, , \tag{9.8}$$

with the use of atomic operators (3.10):

$$\hat{\rho}_A = w_A(\vec{r}\,')\big|_{\vec{r}\,'=\vec{r}} \, . \tag{9.9}$$

Here the atomic operators $\hat{\rho}_A$ are Hermitian, therefore the sum with the adjoints shown in (3.1) is not indicated in (9.8). Furthermore, if no quantum mechanical operators are explicitly considered, then the distinction of primed and unprimed quantities in (9.9) is superfluous, and we may simply write

$$\hat{\rho}_A = w_A(\vec{r}) \, . \tag{9.10}$$

Gross atomic populations

One may follow two slightly different approaches leading, of course, exactly to the same results. We may start from the normalization integral of the electron density and insert the condition (9.3):

$$N = \int \varrho(\vec{r})dv \equiv \int \sum_A w_A(\vec{r})\varrho(\vec{r})dv = \sum_A \int w_A(\vec{r})\varrho(\vec{r})dv = \sum_A Q_A \, . \tag{9.11}$$

Then one gets quite naturally the number of electrons N as a sum of 3D (gross) atomic populations

$$Q_A = \int w_A(\vec{r})\varrho(\vec{r})dv \, . \tag{9.12}$$

Formula (9.12) is valid for both disjunct and "fuzzy" atomic domains. In the case of the former $w_A(\vec{r})$ is either 1 or 0, therefore Q_A can be rewritten as an integral over the respective atomic domain:

$$Q_A^{AIM} = \int_{\Omega_A} \varrho(\vec{r}) dv \, , \tag{9.13}$$

where the superscript indicates that this formula is pertinent to the theories using disjunct atomic domains, the most important representative of which is Bader's AIM theory. That means that in this case the decomposition (9.11) can simply be obtained by writing the integration over the whole 3D space as a sum of integration over the atomic domains Ω_A. In fact, one has

$$\int f(\vec{r}) dv \equiv \sum_A \int_{\Omega_A} f(\vec{r}) dv \, , \tag{9.14}$$

for any function $f(\vec{r})$.

Alternatively, one can express the number of electrons (integral of the electron density) through the elements of the density-matrix and overlap matrix:

$$N = Tr(\mathbf{DS}) = \sum_{\mu,\nu} D_{\mu\nu} S_{\nu\mu} \, , \tag{9.15}$$

and insert in the "ket" of each overlap matrix $S_{\nu\mu} = \langle \chi_\nu | \chi_\mu \rangle$ the atomic resolution of identity (9.8), similarly to that made in Eq. (3.18)

$$S_{\nu\mu} = \langle \chi_\nu | \chi_\mu \rangle = \sum_A \langle \chi_\nu | \hat{\rho}_A \chi_\mu \rangle = \sum_A S_{\nu\mu}^A \, , \tag{9.16}$$

Then one gets for the gross atomic population of atom A

$$Q_A = Tr(\mathbf{DS}^A) = \sum_{\mu,\nu} D_{\mu\nu} S_{\nu\mu}^A \, , \tag{9.17}$$

where $\mathbf{S}^A$ is the "atomic density matrix" with the elements $S_{\nu\mu}^A$. Contrary to the case of Hilbert-space analysis, here all the "atomic overlap integrals" $S_{\nu\mu}^A$ can be expected to differ from zero, because they are defined as

$$S_{\nu\mu}^A = \int w_A(\vec{r}) \chi_\nu^*(\vec{r}) \chi_\mu(\vec{r}) dv \, . \tag{9.18}$$

In the case of Bader's theory, the "atomic overlap integral" can also be expressed as an integral restricted to the respective atomic domain:

$$S_{\nu\mu}^{A(AIM)} = \int_{\Omega_A} \chi_\nu^*(\vec{r}) \chi_\mu(\vec{r}) dv \, . \tag{9.19}$$

Overlap populations

As discussed in Section 3.3, one obtains net atomic populations and overlap populations by applying two atomic resolutions of identity. Note that if disjunct atomic domains are used, like in Bader's AIM theory, then the product of two different atomic operators annihilates any function (the weight functions of two different atoms do not simultaneously differ from zero); consequently in such theories the net and gross atomic populations are equal and the overlap populations vanish identically.

Inserting the decomposition of identity (9.3) into the integral of electronic density in Eq. (9.11) *twice* (using summations indices A and B, respectively), one gets

$$N=\int \varrho(\vec{r})dv \equiv \int \sum_{A,B} w_A(\vec{r})w_B(\vec{r})\varrho(\vec{r})dv \equiv \sum_{A,B}\int w_A(\vec{r})w_B(\vec{r})\varrho(\vec{r})dv = \sum_{A,B} q_{AB} \, , \tag{9.20}$$

where

$$q_{AB} = \int w_A(\vec{r})w_B(\vec{r})\varrho(\vec{r})dv \, . \tag{9.21}$$

Obviously, the diagonal terms q_{AA} (i.e., those for which $A = B$) are the net populations and those q_{AB} for which $A \neq B$ the overlap populations in the "fuzzy atoms" framework. The latter measure the extent to which the atomic charge is shared between atoms A and B, owing to the fact that q_{AB} will have any appreciable value only if there is a part of space in which both weight functions w_A and w_B differ from zero significantly. It is trivial to see that one has the identity

$$Q_A = \sum_B q_{AB} \, . \tag{9.22}$$

One may also apply Eq. (3.22) in order to express q_{AB} in terms of the density matrix. One gets

$$q_{AB} = \sum_{\mu,\nu} D_{\mu\nu} \int w_A(\vec{r})w_B(\vec{r})\chi_\mu^*(\vec{r})\chi_\nu(\vec{r})dv \, . \tag{9.23}$$

Obviously, expressions in terms of electron density can be used for both LCAO-type calculations (use of atom-centered basis sets) and plane wave calculations, those in terms of density matrices can directly be used only in the LCAO ones.

The results in Table 9.1 illustrate well that the populations obtained in the 3D analysis indeed converge quickly as the basis set improves. They do, however, depend on the exact definition of the atoms in the 3D space—if Becke atoms are used, then on the parameter k of the procedure (number of iterations, see Appendix A.2) and on the atomic radii used.

Table 9.1
Basis set dependence of net atomic populations and overlap populations calculated by the "fuzzy atoms" formalism and Becke's weigh functions

Ethane

Basis	C	H	C–C	C–H
STO-3G	5.439	0.809	0.285	0.136
6-31G	5.433	0.810	0.282	0.136
6-31G(d,p)	5.412	0.808	0.286	0.140
6-311G(d,p)	5.409	0.809	0.285	0.141
6-311++G(d,p)	5.409	0.809	0.285	0.141
cc-pVDZ	5.409	0.809	0.284	0.141
cc-pVTZ[a]	5.405	0.808	0.287	0.141
aug-cc-pVDZ	5.410	0.808	0.285	0.141
6-311++G(3df,pd)[a]	5.405	0.808	0.286	0.141

Water

Basis	O	H	O–H
STO-3G	7.816	0.761	0.163
6-31G	7.852	0.713	0.179
6-31G(d,p)	7.789	0.722	0.190
6-311G(d,p)	7.783	0.725	0.189
6-311++G(d,p)	7.781	0.723	0.191
cc-pVDZ	7.779	0.727	0.190
cc-pVTZ[a]	7.768	0.725	0.194
aug-cc-pVDZ	7.767	0.726	0.193
6-311++G(3df,pd)[a]	7.770	0.723	0.194

[a]Using 10 f-orbitals.
*[122] with permission from Elsevier.

9.3 Bond orders and valences from the 3D space analysis

One may start from the expression (3.43) of the exchange density, and its integral (3.44). Obviously, these formulae may be rewritten as

$$N = \sum_{\mu,\nu,\rho,\tau} \left[P^{\alpha}_{\mu\nu} \int \chi^{\star}_{\nu}(\vec{r}_1)\chi_{\rho}(\vec{r}_1)dv_1 P^{\alpha}_{\rho\tau} \int \chi^{\star}_{\tau}(\vec{r}_2)\chi_{\mu}(\vec{r}_2)dv_2 \right. \tag{9.24}$$
$$\left. + P^{\beta}_{\mu\nu} \int \chi^{\star}_{\nu}(\vec{r}_1)\chi_{\rho}(\vec{r}_1)dv_1 P^{\beta}_{\rho\tau} \int \chi^{\star}_{\tau}(\vec{r}_2)\chi_{\mu}(\vec{r}_2)dv_2 \right]$$

The use of this formula is obvious for an LCAO calculation; the problem of applying the result obtained in the plane wave case will be discussed later. (Of course, one can always turn to the use of MO basis and apply the formulae obtained for plane waves also if atom-centered basis sets are used.)

Now we insert identity (9.3) in each integral on the right-hand side of (9.24), introduce the "atomic overlap matrices" $\mathbf{S}^A$, $\mathbf{S}^B$ with the elements defined in Eq. (9.18), and get [122]:

$$N = \sum_{A,B}\sum_{\mu,\rho}\left[(\mathbf{P}^\alpha\mathbf{S}^A)_{\mu\rho}(\mathbf{P}^\alpha\mathbf{S}^B)_{\rho\mu} + (\mathbf{P}^\beta\mathbf{S}^A)_{\mu\rho}(\mathbf{P}^\beta\mathbf{S}^B)_{\rho\mu}\right] \tag{9.25}$$

Similarly to Section 6.3, we can turn to the matrices of total density $\mathbf{D}$ and spin-density $\mathbf{P}^s$ defined in Eq. (6.16), and get (9.25) in the form

$$2\sum_A Q_A = \sum_{A,B}\sum_{\mu,\rho}\left[(\mathbf{D}\mathbf{S}^A)_{\mu\rho}(\mathbf{D}\mathbf{S}^B)_{\rho\mu} + (\mathbf{P}^s\mathbf{S}^A)_{\mu\rho}(\mathbf{P}^s\mathbf{S}^B)_{\rho\mu}\right] \tag{9.26}$$

where the equality $\sum_A Q_A = N$ has also been utilized.

The *bond order* between atoms A and B is, in general, defined as the diatomic contribution to the normalization of the exchange density (see Section 6.3 and [63, 64]). Thus, in the 3D case we arrive at the definition [122,126]:

$$\begin{aligned} B_{AB} &= \sum_{\mu,\rho}\left[(\mathbf{D}\mathbf{S}^A)_{\mu\rho}(\mathbf{D}\mathbf{S}^B)_{\rho\mu} + (\mathbf{P}^s\mathbf{S}^A)_{\mu\rho}(\mathbf{P}^s\mathbf{S}^B)_{\rho\mu}\right] \\ &\equiv 2\sum_{\mu,\rho}\left[(\mathbf{P}^\alpha\mathbf{S}^A)_{\mu\rho}(\mathbf{P}^\alpha\mathbf{S}^B)_{\rho\mu} + (\mathbf{P}^\beta\mathbf{S}^A)_{\mu\rho}(\mathbf{P}^\beta\mathbf{S}^B)_{\rho\mu}\right] \end{aligned} \tag{9.27}$$

This equation can be considered the most general definition of bond order for single determinant wave functions: it is valid not only for different versions of 3D analysis* [122,126], but also covers the case of Hilbert-space analysis, if one considers the atomic overlap matrices with the "vertical band structure" discussed on page 24. (As will be discussed later, even the bond

*For the AIM framework this bond order formula was introduced in Ref. 126. Somewhat later it was renamed "delocalization index" by the Bader group. Thus for the single determinant (Hartree–Fock, DFT) wave functions "bond order" and "delocalization index" mean the same. However, for *correlated* wave function people use the term "delocalization index" for a quantity which can be related to the "fluctuation definition" Eq. (6.40) and, therefore, differs from the quantity we suggest to use in the correlated case. (Its calculation requires the explicit use of the second-order density matrix, while for our definition one needs only the first-order one.) It is to be noted here that the recent improvement of the bond order calculation in the correlated case, using matrix $\mathbf{R}$, described in Section 6.7, has not yet been implemented in the framework of the 3D analysis, but that should not represent a difficulty.

orders corresponding to the results of plane wave calculations can be determined using this formula.)

The valence and free valence indices in the 3D framework are defined quite analogously to the case of Hilbert-space analysis (Section 6.10) [122,126].

The *total valence of atom A* is defined as

$$V_A = 2Q_A - \sum_{\mu,\rho} (\mathbf{DS}^A)_{\mu\rho} (\mathbf{DS}^A)_{\rho\mu} \ , \tag{9.28}$$

and the *free valence of atom A* as the difference

$$F_A = V_A - \sum_{\substack{B \\ B \neq A}} B_{AB} \ . \tag{9.29}$$

Here Q_A, of course, means the gross atomic population Eq. (9.12) determined for the actual (AIM of "fuzzy atoms") 3D formalism.

Similar to the Hilbert-space analysis, in the single determinant (UHF) case the free-valence index F_A can also be expressed through the spin-density matrix:

$$F_A = \sum_{\mu,\rho} (\mathbf{P}^s\mathbf{S}^A)_{\mu\rho} (\mathbf{P}^s\mathbf{S}^A)_{\rho\mu} \qquad \text{(UHF)} \ . \tag{9.30}$$

It vanishes for the closed shell RHF case for which the spin density is zero. In the RHF case—similar to its Hilbert-space counterpart—the total valence of an atom equals the sum of all its bond orders:

$$V_A = \sum_{\substack{B \\ B \neq A}} B_{AB} \qquad \text{(RHF)} \ . \tag{9.31}$$

Table 9.2 contains some selected valence and bond order values [122] calculated by using Becke's weight functions and two different basis sets—one of which contains diffuse functions. The results indicate that bond orders and valences calculated in the framework of 3D analysis—similar to the results of population analysis—exhibit only a rather small basis dependence. An interesting observation is that valences obtained for Becke atoms are usually somewhat higher than the classical integer values, which is in contrast with the Hilbert space results: for the latter, if good atomic type basis sets are used then one gets numbers close to the ideal ones, but mostly remain somewhat below them.

The definition of bond orders and valences for the plane wave calculations required a special approach [127]. These calculations have been developed in recent decades for treating special (often rather large) systems, by applying

Table 9.2
Valences and bond orders calculated by the "fuzzy atoms" formalism and Becke's weight function
Basis sets: 6-31G** (A), and 6-311++G** (B).

		Valences			Bond orders	
		A	B		A	B
H_2	H	1.000	1.000	H-H	1.000	1.000
N_2	N	3.103	3.109	N-N	3.103	3.109
HF	H	0.899	0.914	H-F	0.899	0.914
	F	0.899	0.914			
CO	C	2.766	2.779	C-O	2.766	2.779
	O	2.766	2.779			
H_2O	O	2.337	2.368	O-H	1.169	1.184
	H	1.236	1.254			
NH_3	N	3.218	3.256	N-H	1.073	1.085
	H	1.186	1.200			
B_2H_6	B	3.719	3.718	B-H_{br}	0.460	0.460
	H_{br}	0.996	0.996	B-H_t	0.943	0.943
	H_t	1.022	1.023	B-B	0.846	0.845
SO	S	2.622	2.614	S-O	2.622	2.614
	O	2.622	2.614			
SO_2	S	4.979	4.959	S-O	2.490	2.480
	O	2.694	2.690			
SO_3	S	7.086	7.045	S-O	2.362	2.348
	O	2.633	2.634			
CH_4	C	3.939	3.937	C-H	0.985	0.984
	H	1.115	1.116			
C_2H_6	C	4.151	4.151	C-C	1.130	1.128
	H	1.103	1.105	C-H	0.951	0.951
C_2H_4	C	4.056	4.054	C-C	1.976	1.960
	H	1.094	1.101	C-H	0.963	0.967
C_2H_2	C	3.934	3.935	C-C	2.865	2.856
	H	1.072	1.082	C-H	0.986	0.991
C_6H_6	C	4.252	4.255	C-C	1.440	1.436
	H	1.078	1.086	C-H	0.937	0.940
C_{60}	C	4.35[a]		C-C(6,6)	1.42[a]	
				C-C(5,6)	1.13[a]	

[a]Single point, 6-31G basis, smaller integration grid. (No symmetry is utilized.)
*[122] with permission of Elsevier.

the formalism of an infinite number of repeated subunits with periodic boundary conditions. In these calculations one uses plane waves as basis orbitals, and not atom-centered basis functions that are more customary for quantum chemists. Special care should be taken to avoid a spurious "interaction" between the repeated units. In these calculations one usually applies a DFT formalism and effective cores; that means that one essentially needs to calculate the valence charge density in different points of the elementary cell in order to define the effective Kohn–Sham Fockian. Direct use of Eqs. (9.27)–(9.29) for the plane wave basis functions seems hopeless. However, these formulae are applicable for any basis set in which the wave function can be expressed—thus also for the canonical molecular orbitals. For an RHF or ROHF calculation (double filled orbitals plus possibly some single occupied orbitals of α spin) the density matrices $\mathbf{D}$ and $\mathbf{P}^s$ are diagonal in terms of the MOs with diagonal elements equal to 1 or 2, and 0 or 1, respectively. The "atomic density matrices" should also be calculated over the MOs, i.e., their elements are

$$S_{ij}^{A} = \int w_A(\vec{r})\varphi_i^{\star}(\vec{r})\varphi_j(\vec{r})dv \quad , \tag{9.32}$$

Fortunately, the values of the molecular orbitals can be obtained over a grid defined in the elementary cell, so the integration in Eq. (9.32) can be performed numerically. In this case the bond order formula (9.27) reduces to

$$B_{AB} = \sum_{i,j} n_i n_j S_{ij}^{A} S_{ji}^{B} + \sum_{i,j}^{s.occ.} S_{ij}^{A} S_{ji}^{B} \quad , \tag{9.33}$$

where n_i is the occupation number of the MO φ_i and "$s.occ.$" stands for "singly occupied."

Analogously, valences are given by Eq. (9.31) for a closed-shell calculation, or by

$$V_A = 2Q_A - \sum_{i,j} (\mathbf{DS}^A)_{ij} (\mathbf{DS}^A)_{ji} \; , \tag{9.34}$$

in the general case. Free valences can then be determined by using Eq. (9.29).

Of course, Equations (9.32)–(9.34) can be equally well used—and even may appear preferable, especially for large systems—also if one uses a usual LCAO basis set.

In the paper [127] a large number of molecules were considered by using plane wave calculations, applying both Hirshfeld's and Becke's definitions of the atomic weight functions (and different sets of atomic radii for the latter). The numbers are very much similar to those shown in Table 9.2 We shall not display them here, only stress the conceptual importance of this result: quite "chemical" bond order and valence indices are obtained *without introducing any atom-centered basis orbitals* in the calculations. This underscores the correctness of our basic qualitative notions about the electronic structure of molecules; that gets a further confirmation in the results of the next section.

9.4 Effective AOs from the 3D space analysis

As discussed in the paper [42], one can (at least formally) obtain effective atomic orbitals by using *any* bilinear Hermitian localization functional, when forming the Rayleigh quotient (5.8); the possibility to use a "switch-off" weight function of a 3D analysis has also already been mentioned. One could think that in the "fuzzy" atom case one could use simply $w_A(\vec{r})$ for each atom for the definition of the Q-matrix elements, i.e., $Q_{ij} = \langle \varphi_i | w_A(\vec{r}) | \varphi_j \rangle$. However, the numerical experience gave discouraging results: there was no sharp separation between effective minimal basis orbitals and the weakly occupied ones. Then we have realized that the essence of the procedure based on the Rayleigh quotient (5.8) is that the weight of the *intraatomic* part $\hat{\rho}_A \psi$ of the localized orbitals is to be made stationary. The interatomic part $\hat{\rho}_A \psi$ of the function is in the 3D case simply $\psi_A(\vec{r}) = w_A(\vec{r})\psi_A(\vec{r})$, which means that one has to use the *square* of the weight function $w_A(\vec{r})$, that is one should define $Q_{ij} = \langle \varphi_i | w_A^2(\vec{r}) | \varphi_j \rangle$.

These consideration lead to the conclusion that one can obtain effective atomic orbitals in the framework of 3D analysis—either for Bader's disjunct atomic domains or for "fuzzy atoms"—by using the same equations Eq.s (5.7)–(5.11), simply by substituting the definition (5.9) $\hat{Q} = \hat{\rho}_A^\dagger \hat{\rho}_A$ with the pertinent operator $\hat{\rho}_A$ as defined in Eq. (3.10) and the respective weight functions. That will result in elements of matrix **Q** which are simply equal to

$$Q_{ij}^{AIM} = \langle \varphi_i | \hat{Q}^{AIM} | \varphi_j \rangle = \int_{\Omega_A} \varphi_i^*(\vec{r})\varphi_j(\vec{r}) dv \qquad (9.35)$$

in the Bader case (Ω_A being the domain of atom A) and

$$Q_{ij}^{fuzzy} = \langle \varphi_i | \hat{Q}^{fuzzy} | \varphi_j \rangle = \int w_A^2(\vec{r})\varphi_i^*(\vec{r})\varphi_j(\vec{r}) dv \qquad (9.36)$$

in the "fuzzy" atom one.* The definitions (9.35) and (9.36) are completely consistent with the Hilbert space analysis discussed in Section 5.1, and similar to that case permit to define *orthonormalized atomic hybrids* as

$$\psi_i^A = (M_i)^{-1/2} \hat{\rho}_A \varphi_i' = w_A(\vec{r}) \varphi_i'(\vec{r}) \; , \qquad (9.37)$$

*In both 3D cases the atomic operators $\hat{\rho}_A$ are Hermitian, in the Bader case also idempotent.

where φ_i' is the localized MO obtained after diagonalizing matrix **Q**, as given by Eq. (5.7).

Within this framework, effective AOs have been calculated for both Bader's AIM framework [129] and by using the "fuzzy atoms" formalism. In the latter case the formalism has been applied to calculations performed not only by using atom-centered basis sets [130] but also for plane-wave basis [128]. In every case similar behavior has been observed to that discussed in Section 5.2 for the effective AOs calculated by using Hilbert-space analysis: there are as many strongly occupied orbitals as the number of functions in the classical minimal basis. Also the behavior of the hypervalent systems is quite similar to that discussed in the Hilbert space analysis.

We attribute extraordinary importance to the fact that functions strongly resembling the ordinary atomic orbitals (only being restricted to the atomic domains) have been obtained from the plane-wave calculations *without introducing any atom-centered basis orbitals* in the procedure. This is illustrated by Figures 9.2–9.5. These observations permit one to conclude that the existence of effective minimal basis sets featuring *s*, *p*, *d*-type orbitals or their hybrids is a "law of nature" and not an artifact of the computations. Thus the results provide a very important confirmation for the basis ideas of our qualitative picture of molecular electronic structure.

9.5 Interrelation between Hilbert-space and 3D analyses

In this section we are going to consider some results of *"Hilbert-space"* analysis performed by using the effective AOs obtained in a fuzzy atoms analysis as basis orbitals. Following the considerations in [131], we shall prove that the Mulliken net and overlap populations *calculated in that special basis* are equal to the "fuzzy atoms" net and overlap populations. The same holds also for the bond orders calculated in that basis and in the "fuzzy atoms" framework. We attribute big conceptual importance to these results, owing to the fact that "Hilbert-space" analysis and "fuzzy atoms' analysis differ in significant aspects.

Electron populations

Every molecular orbital can be presented in terms of its intraatomic com-

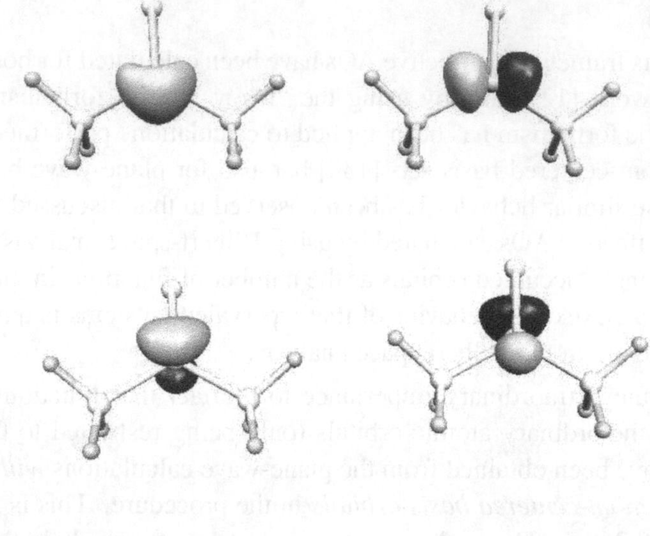

Figure 9.2

Strongly occupied valence orbitals of the central carbon atom in the acetone molecule as extracted from plane wave calculations. [128] with permission from American Chemical Society.

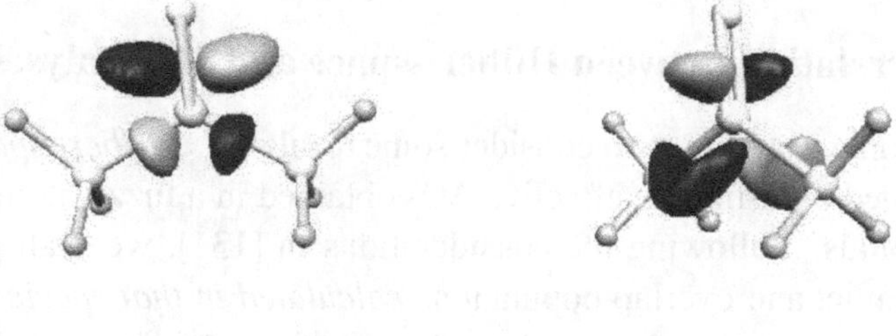

Figure 9.3

First two weakly occupied orbitals of the central carbon atom in the acetone molecule as extracted from plane wave calculations. [128] with permission from American Chemical Society.

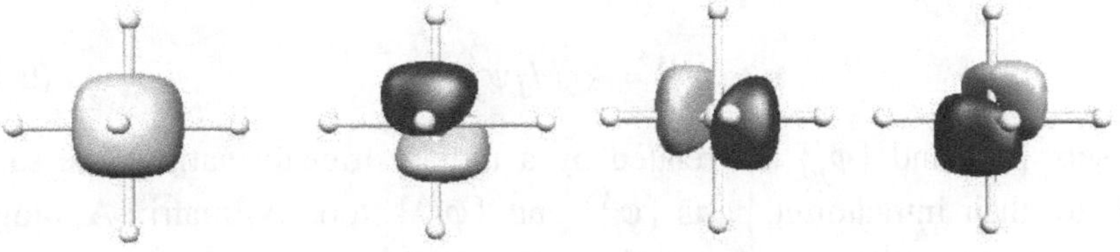

Figure 9.4
Strongly occupied valence orbitals of the sulfur atom in the SF_6 molecule as extracted from plane wave calculations. [128] with permission from American Chemical Society.

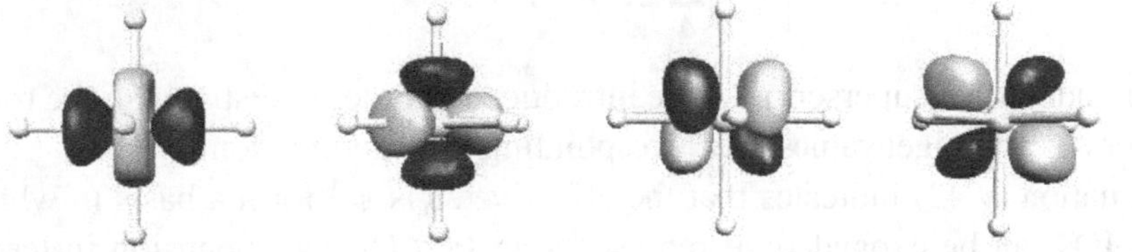

Figure 9.5
First weakly occupied valence orbitals of the sulfur atom in the SF_6 molecule as extracted from plane wave calculations. [128] with permission from American Chemical Society.

ponents (5.4) corresponding to the different atoms:

$$\varphi_i(\vec{r}) = \sum_A \varphi_i^A(\vec{r}) = \sum_A w_A(\vec{r})\varphi_i(\vec{r}) , \tag{9.38}$$

owing to the equality $\sum_A w_A(\vec{r}) = 1$, valid in every point of space. This is true for both the original (say canonical) orbitals φ_i and for those φ_j'-s that have been obtained when searching for the stationary value of the Rayleigh quotient (5.8) for the given atom, i.e., solving the eigenvalue equations (5.11). According to (5.16), the *intraatomic parts*

$$\varphi_j'^A(\vec{r}) = w_A(\vec{r})\varphi_j'(\vec{r})a , \tag{9.39}$$

corresponding to non-zero norms M_j, give rise after renormalization to the effective AOs ψ_j^A. Therefore, one has

$$\varphi_j'^A = \sqrt{M_j}\psi_j^A \,. \tag{9.40}$$

The sets $\{\varphi_i\}$ and $\{\varphi_j'\}$ are related by a unitary transformation; the same holds for their intraatomic parts $\{\varphi_i^A\}$ and $\{\varphi_j'^A\}$, too. As matrix **A**, diagonalizing the Hermitian matrix is unitary, $\mathbf{A}^{-1} = \mathbf{A}^\dagger$, one has

$$\varphi_i^A = \sum_j a_i^{j*}\varphi_j'^A = \sum_j \sqrt{M_j}a_i^{j*}\psi_j^A a \,. \tag{9.41}$$

Therefore, one can express the molecular orbitals as linear combinations of the effective AOs of all the atoms:

$$\varphi_i = \sum_A \sum_j \sqrt{M_j^A a_i^{Aj*}}\,\psi_j^A \,, \tag{9.42}$$

where additional superscripts A are introduced in order to distinguish the row-vectors $\mathbf{a}_i^\dagger$ and eigenvalues M_i corresponding to different atoms.

Equation (9.42) indicates that the effective AOs ψ_j^A form a basis in which the MOs can be expanded: it represents an LCAO-type expansion in terms of the effective AOs.* In the present section we shall understand term LCAO right in that sense. To stress that we are dealing with the LCAO expression in terms of effective atomic orbitals, the respective quantities will be denoted LCEAO ("E" for *effective*).

The LCAO coefficient of the basis orbital ψ_j^A of the MO φ_i in the expansion (9.42) is

$$C_{ji}^A = \sqrt{M_j^A a_i^{Aj*}} \,. \tag{9.43}$$

The density matrix element then becomes

$$D_{jk}^{AB} = 2\sum_i^{occ.} C_{ji}^A C_{ki}^{B*} = 2\sum_i^{occ.} \sqrt{M_j^A a_i^{Aj*}}\sqrt{M_k^B a_i^{Bk}} \,. \tag{9.44}$$

The Mulliken's net and overlap populations (depending on whether $A = B$ or $A \neq B$) are, according to the expressions (4.26)

$$q_{AB}^{LCEAO} = \sum_{\mu\in A}\sum_{\nu\in B} D_{\mu\nu}S_{\nu\mu} = 2\sum_{j,k}\sum_i^{occ.} \sqrt{M_j^A a_i^{Aj*}}\sqrt{M_k^B a_i^{Bk}}\langle\psi_k^B|\psi_j^A\rangle \,, \tag{9.45}$$

---

* Obviously, only effective AOs with the eigenvalues $M_j^A > 0$ contribute to the expansion, but the conservation of the terms corresponding to the zero M_j^A values (if any) shall simplify our following considerations. In the case of $M_i^A = 0$, the corresponding ψ_j^A may be assumed to be any *finite* function.

where the explicit expression of the overlap integral of the effective atomic orbitals $\langle \psi_k^B | \psi_j^A \rangle$ has also be written out.

The fuzzy atom net or overlap populations are given by Eq. (9.21). Substituting here the electronic density (1.3), one gets

$$q_{AB}^{fuzzy} = \int w_A(\vec{r}) w_B(\vec{r}) \varrho(\vec{r}) dv = \int w_A(\vec{r}) w_B(\vec{r}) 2 \sum_i^{occ.} \varphi_i^*(\vec{r}) \varphi_i(\vec{r}) dv \,. \quad (9.46)$$

This can also be written as

$$q_{AB}^{fuzzy} = 2 \sum_i^{occ.} \langle w_A(\vec{r}) \varphi_i(\vec{r}) | w_B(\vec{r}) \varphi_i(\vec{r}) \rangle \,. \quad (9.47)$$

It follows from the relationship (5.7) describing the unitary transformation diagonalizing matrix $\mathbf{Q}$ that for every atom

$$\varphi_i = \sum_j a_i^{Aj*} \varphi_j' \,. \quad (9.48)$$

Comparing relationships (9.39), (9.40) we have $\varphi_j^{\prime A}(\vec{r}) = \sqrt{M_j^A} \psi_j^A(\vec{r})$. so one can further write

$$w_A(\vec{r}) \varphi_i(\vec{r}) = \sum_j a_i^{Aj*} w_A(\vec{r}) \varphi_j'(\vec{r}) = \sum_j a_i^{Aj*} \varphi_j^{\prime A}(\vec{r}) = \sum_j a_i^{Aj*} \sqrt{M_j^A} \psi_j^A(\vec{r}) \,, \quad (9.49)$$

and similarly for atom B:

$$w_B(\vec{r}) \varphi_i(\vec{r}) = \sum_k a_i^{Bk*} \sqrt{M_k^B} \psi_k^B(\vec{r}) \,. \quad (9.50)$$

Substituting expressions (9.49) and (9.50) into (9.47), one gets

$$q_{AB}^{fuzzy} = 2 \sum_i^{occ.} \sum_{j,k} a_i^{Aj*} \sqrt{M_j^A} a_i^{Bk} \sqrt{M_k^B} \langle \psi_k^B | \psi_j^A \rangle \,. \quad (9.51)$$

Comparing this result with Eq. (9.45), one sees that

$$q_{AB}^{LCEAO} = q_{AB}^{fuzzy} \,. \quad (9.52)$$

Gross atomic populations are sums of the net and overlap ones:

$$Q_A = \sum_B q_{AB} = q_{AA} + \sum_{\substack{B \\ B \neq A}} q_{AB} \,, \quad (9.53)$$

therefore the same holds for the *gross atomic populations*, too:

$$Q_A^{LCEAO} = Q_A^{fuzzy} \ . \tag{9.54}$$

Similar relationship holds [129] in the Bader-type 3D theory; it may be considered a consequence of the fact that Bader's AIM results represent a limiting case of "fuzzy atoms" analysis with $w_A(\vec{r})$ values equal to either 1 or 0.

Bond orders

According to Eq. (6.18), the Hilbert-space bond order index between atoms A and B is defined in the closed-shell case as

$$B_{AB} = \sum_{\mu \in A} \sum_{\nu \in B} (\mathbf{DS})_{\mu\nu} (\mathbf{DS})_{\nu\mu} \ . \tag{9.55}$$

Using the expression (9.44) of the density matrix elements, this can be rewritten as

$$\begin{aligned} B_{AB}^{LCEAO} &= 4 \sum_{i,j}^{occ.} \sum_{C,D} \sum_{k,l,p,q} C_{ki}^{A} C_{pi}^{C*} S_{pl}^{CB} C_{lj}^{B} C_{qj}^{D*} S_{qk}^{DA} \\ &= 4 \sum_{i,j}^{occ.} \sum_{k,l,p,q} \langle \sum_{C} \sum_{p} C_{pi}^{C} \psi_p^{C} | \psi_l^{B} \rangle C_{lj}^{B} \langle \sum_{D} \sum_{q} C_{qj}^{D} \psi_q^{D} | \psi_k^{A} \rangle C_{ki}^{A} \ . \end{aligned} \tag{9.56}$$

We observe, that the "bras" contain the expansions of the canonical molecular orbitals φ_i and φ_j, respectively. Furthermore, it follows from Eqs. (9.39), (9.40) that*

$$\psi_j^A = \frac{1}{\sqrt{M_j^A}} w_A(\vec{r}) \varphi_j^{[A]\prime}(\vec{r}) \ , \tag{9.57}$$

where the additional superscript "[A]" has been introduced to stress that $\varphi_j^{[A]\prime}(\vec{r})$ has been obtained by solving the eigenvalue equation of matrix $\mathbf{Q}$ built up for atom A. (The square braces in the superscript are introduced to distinguish the MO $\varphi_j^{[A]\prime}(\vec{r})$ from the intraatomic part $\varphi_j^{A\prime}(\vec{r})$ of the MO $\varphi_j^{\prime}(\vec{r})$, that was used above.) Substituting this, we get

$$B_{AB}^{LCEAO} = 4 \sum_{i,j}^{occ.} \sum_{k,l} \langle \varphi_i | \frac{1}{\sqrt{M_l^B}} w_B(\vec{r}) \varphi_l^{[B]\prime} \rangle C_{lj}^{B} \langle \varphi_j | \frac{1}{\sqrt{M_k^A}} w_A(\vec{r}) \varphi_k^{[A]\prime} \rangle C_{ki}^{A} \ . \tag{9.58}$$

*In accordance with the footnote on page 194, only ψ_j^As corresponding to non-zero M_j^A values should be accounted for in the expansions.

Substituting here C^B_{lj} and C^A_{ki} according to (9.43), and utilizing (9.48), this simplifies to

$$B^{LCEAO}_{AB} = 4 \sum_{i,j}^{occ.} \langle \varphi_i | w_B(\vec{r}) | \varphi_j \rangle \langle \varphi_j | w_A(\vec{r}) | \varphi_i \rangle = 4 \sum_{i,j}^{occ.} S^B_{ij} S^A_{ji} \,. \tag{9.59}$$

Here S^B_{ij} and S^A_{ji} are the atomic overlap integrals over the MOs, defined in Eq. (9.32).

The bond order for the fuzzy atom analysis can be obtained by using Eq. (9.33), taking into account that we are dealing with a closed-shell case, so the terms corresponding to singly occupied orbitals should be omitted and all occupation numbers $n_i = n_j = 2$. Thus we have

$$B^{fuzzy}_{AB} = 4 \sum_{i,j}^{occ.} S^A_{ij} S^B_{ji} \,. \tag{9.60}$$

Comparing expressions (9.59) and (9.60), we arrive at

$$B^{LCEAO}_{AB} = B^{fuzzy}_{AB} \,. \tag{9.61}$$

As noted above, we consider the equalities (9.52), (9.54), and (9.61) to be of great conceptual importance. Also, they can be used as an additional argument for our old opinion about the mathematical importance of Mulliken's populations, discussed in Section 3.3: these results show that if there occur problems with the Mulliken-type analysis, then their origin is not in the formalism but in the use of an inadequate basis set.

9.6 Energy components in the 3D space

Energy components in the AIM theory

In a theory in which the 3D space is decomposed into *disjunct* atomic domains Ω_A, one has identity (9.14) for any one-electron integral $\int f(\vec{r})dv$, i.e., any one-electron integral spontaneously decomposes into a sum of atomic contributions. Analogously, one has that any two-electron-integral is a trivial sum of one- and two-center components (the former occurs in the case if $A = B$):

$$\int g(\vec{r}_1, \vec{r}_2) dv_1 dv_2 \equiv \sum_{A,B} \int_{\Omega_A} \int_{\Omega_B} g(\vec{r}_1, \vec{r}_2) dv_1 dv_2 \,, \tag{9.62}$$

for any function $g(\vec{r}_1,\vec{r}_2)$. Similar to (9.14), the identity (9.62) may also be obtained formally by inserting the atomic decomposition of identity (9.3) with the definition (9.4) on the left-hand side, but twice, not just once.

The SCF energy contains only one- and two-electron integrals; the identities (9.62) and (9.62) indicate that it spontaneously decomposes [112] into a sum of atomic and diatomic contributions in the AIM framework and for any theories in which the 3D space is decomposed into disjunct atomic domains.* Thus, for instance, the RHF energy becomes trivially

$$E = \sum_A E_A + \sum_{A<B} E_{AB} \ , \tag{9.63}$$

where

$$E_A = 2\sum_i^{occ.} \langle\varphi_i|\hat{h}^A|\varphi_i\rangle_A + \sum_{i,j}^{occ.}\left(2[\varphi_i\varphi_j|\varphi_i\varphi_j]_{A,A} - [\varphi_i\varphi_j|\varphi_j\varphi_i]_{A,A}\right) \ ; \tag{9.64}$$

$$E_{AB} = -2\sum_i^{occ.}\left(\langle\varphi_i|\frac{Z_B}{R_B}|\varphi_i\rangle_A + \langle\varphi_i|\frac{Z_A}{R_A}|\varphi_i\rangle_B\right)$$
$$+2\sum_{i,j}^{occ.}\left(2[\varphi_i\varphi_j|\varphi_i\varphi_j]_{A,B} - [\varphi_i\varphi_j|\varphi_j\varphi_i]_{A,B}\right) + \frac{Z_A Z_B}{R_{AB}} \ . \tag{9.65}$$

Here, as usual

$$\hat{h}^A = -\tfrac{1}{2}\Delta - \frac{Z_A}{r_A} \ , \tag{9.66}$$

is the one-electron *atomic* Hamiltonian for atom A and the subscripts indicate the domains of integration.

This result [112] has a big conceptual importance, because it shows that the HF energy of a molecule can *exactly* be decomposed into the sum of atomic and diatomic contributions in the 3D space analysis. (That is also true not only if the 3D space is decomposed into disjunct atomic domains but also if fuzzy ones are considered, as will be described in the next section.)

Using the method of simple "mapping" between the "3D analysis" and "Hilbert space analysis," one could derive [112] the energy components of the CECA method of Section 8.4 starting from Eqs. (9.64), (9.65). As already noted, this may be considered an additional theoretical justification of the CECA scheme.

---

*It is to be noted that this energy decomposition has nothing in common with that based on the "local virial theorem" of the AIM theory, in which the total energy is presented (approximately) as a sum of atomic energies only.

The actual application of the energy decomposition with the components Eqs. (9.64), (9.65) requires performing numerical integrations which are complicated by the fact that the AIM atomic domains often have complex form.* Such a task has been accomplished in [132]. Table 9.3 displays some results obtained, It can be seen that the numbers are qualitatively similar to those obtained by the CECA "Hilbert space" method, in agreement with the conceptual connection between them, mentioned above.

This table also represents an example of the greatest conceptual difficulty occurring in Bader's AIM method: for any basis set containing polarization functions, there is a "non-nuclear attractor" in the center of the C≡C bond of the acetylene molecule, that obviously makes no chemical sense. (For that reason the acetylene data in the table refer to the 6-31G basis lacking polarization functions.)

Table 9.3
AIM energy components of the ethane, ethylene, and acetylene molecules obtained by using 6-31G** (ethane, ethylene) and 6-31G (acetylene) basis sets

Atom(s)	Energy component (a.u.)		
	Ethane	Ethylene	Acetylene
C	-37.3567	-37.3767	-37.4304
H	-0.4330	-0.4339	-0.4298
C–C	-0.2625	-0.4567	-0.6061
C–H	-0.2608	-0.2586	-0.2346
H–H geminal	-0.0051	-0.0041	—
H–H vicinal	0.0007 (2x)	-0.0004	
	0.0003 (1x)	-0.0008	
C–H vicinal	-0.0088	-0.0073	-0.0057
Sum of components	-79.2087	-77.9911	-76.8026
Exact HF energy	-79.2382	-78.0388	-76.7927

*[132] with permission from American Institute of Physics.

The above result of [112] has been "expropriated" in some sense by a Spanish group who renamed the energy decomposition into the method of "inter-

*Owing to the complexity of the domains, special care should be taken when performing the numerical integrations in order to avoid significant round-off errors.

acting quantum atoms"; the respective priority problem has been clarified at a conference [133].

Energy components for "fuzzy atoms"

The energy decomposition Eqs. (9.63)–(9.65) may be considered very "natural" because one obtains the formulae by considering the obvious decomposition of every integral into integrals over the individual atomic domains (pairs of domains). One obtains a similar "natural" decomposition if one introduces the decomposition of identity $\sum_A w_A(\vec{r}) = 1$ into each integral. Grouping the terms according to atoms, we essentially recover Eqs. (9.64), (9.65), only the integration over the atomic domain Ω_A should be replaced by integration over the whole 3D space but with the weight function $w_A(\vec{r})$ introduced. Thus we obtain [134]:

$$
\begin{aligned}
E_A = & \, 2\sum_i^{occ.} \int w_A(\vec{r})\varphi_i^*(\vec{r})\hat{h}^A(\vec{r})\varphi_i(\vec{r})dv \\
& + \sum_{i,j}^{occ.} \Big(2 \iint w_A(\vec{r}_1)w_A(\vec{r}_2)\varphi_i^*(\vec{r}_1)\varphi_j^*(\vec{r}_2)\frac{1}{r_{12}}\varphi_i(\vec{r}_1)\varphi_j(\vec{r}_2)dv_1dv_2 \\
& - \iint w_A(\vec{r}_1)w_A(\vec{r}_2)\varphi_i^*(\vec{r}_1)\varphi_j^*(\vec{r}_2)\frac{1}{r_{12}}\varphi_j(\vec{r}_1)\varphi_i(\vec{r}_2)dv_1dv_2 \Big) \; ;
\end{aligned}
\tag{9.67}
$$

$$
\begin{aligned}
E_{AB} = & -2\sum_i^{occ.} \Big(\int w_A(\vec{r})|\varphi_i(\vec{r})|^2 \frac{Z_B}{r_B} dv + \int w_B(\vec{r})|\varphi_i(\vec{r})|^2 \frac{Z_A}{r_A} dv \Big) \\
& + 2\sum_{i,j}^{occ.} \Big(2 \iint w_A(\vec{r}_1)w_B(\vec{r}_2)\varphi_i^*(\vec{r}_1)\varphi_j^*(\vec{r}_2)\frac{1}{r_{12}}\varphi_i(\vec{r}_1)\varphi_j(\vec{r}_2)dv_1dv_2 \\
& - \iint w_A(\vec{r}_1)w_B(\vec{r}_2)\varphi_i^*(\vec{r}_1)\varphi_j^*(\vec{r}_2)\frac{1}{r_{12}}\varphi_j(\vec{r}_1)\varphi_i(\vec{r}_2)dv_1dv_2 \Big) + \frac{Z_A Z_B}{R_{AB}} \, .
\end{aligned}
\tag{9.68}
$$

Table 9.4 contains some numerical results obtained [134] with such a decomposition. It is to be noted here that in the "fuzzy atoms" analysis one can also assign the two-center kinetic energy components to the diatomic contributions, as has been discussed in Section 8.4.5. The possibility to get more "chemical" numbers as well as the inconsistencies met are the same as in the Hilbert-space analogue; we are not going to consider these results in any detail.

Table 9.4
One and two-center energy components (a.u.) calculated by the "fuzzy atoms" formalism and Becke's weight function (6-31G** basis set)

Molecule	Atom	E(A)	Atomic pair	E(A,B)
H_2	H	-0.4477	H–H	-0.2360
N_2	N	-53.9935	N–N	-0.9569
HF	H	-0.3679	H–F	-0.3163
	F	-99.3293		
CO	C	-37.4171	C–O	-0.9550
	O	-74.3691		
SO	S	-397.1004	S–O	-0.9223
	O	-74.2489		
SO_2	S	-396.6782	S–O	-0.9726
	O	-74.2853		
SO_3	S	-396.2293	S–O	-0.9713
	O	-74.3019		
NH_3	N	-53.9834	N–H	-0.3193
	H	-0.4116		
H_2O	O	-74.4748	O–H	-0.3740
	H	-0.3954		
CH_4	C	-37.3747	C–H	-0.2784
	H	-0.4222		
C_2H_6	C	-37.3201	C–C	-0.3912
	H	-0.4161	C–H	-0.2722
C_2H_4	C	-37.3108	C–C	-0.6102
	H	-0.4172	C–H	-0.2736
C_2H_2	C	-37.3109	C–C	-0.8078
	H	-0.4175	C–H	-0.2707
C_6H_6	C	-37.2563	C–C	-0.4803
	H	-0.4128	C–H	-0.2677
B_2H_6	B	-24.3012	B–H_{br}	-0.1573
	H_{br}	-0.3566	B–H_t	-0.2751
	H_t	-0.4106	B–B	-0.1952

*[134] permission from American Institute of Physics.

9.7 Local spins in the 3D space

Following Ref. 135, we shall consider expressions in terms of orthonormalized molecular (natural) orbitals. This is especially convenient if the second-order density matrix, and thus the cumulant, are obtained in their terms.

The spatial distribution of the *number of unpaired electrons*, as given by Eq. (7.11) can be trivially decomposed into atomic components

$$\Delta = \sum_A \Delta_A , \tag{9.69}$$

with

$$\Delta_A = \sum_i \int w_A(\vec{r}) n_i (2 - n_i) \varphi_i^*(\vec{r}) \varphi_i(\vec{r}) dv = \sum_i n_i (2 - n_i) S_{ii}^A , \tag{9.70}$$

where the definition of the "atomic overlap matrices" (9.32) calculated over the MOs (natural orbitals) is utilized.

The expressions for the local spins in the 3D framework can be best obtained by performing the 3D "mapping" of the Hilbert-space formulae Eq. (7.54) and (7.55), i.e., replacing the "atomic overlap matrices" of the Hilbert space analysis, having the "vertical band structure" as discussed on page 24, by the 3D "atomic overlap matrices" (9.32). We get*

$$\begin{aligned}\langle \hat{S}^2 \rangle_A = {} & \frac{3}{4}\Delta_A + \frac{1}{4}(p_A^s)^2 - \frac{1}{4}\sum_{i,j} (\mathbf{P}^s\mathbf{S}^A)_{ij} (\mathbf{P}^s\mathbf{S}^A)_{ji} \\ & + \frac{1}{2}\sum_{i,j,k,l} \left[\Lambda_{ijkl} - \Lambda_{ijlk}\right] S_{ki}^A S_{lj}^A ,\end{aligned} \tag{9.71}$$

and

$$\langle \hat{S}^2 \rangle_{AB} = \frac{1}{4} p_A^s p_B^s - \frac{1}{4}\sum_{i,j} (\mathbf{P}^s\mathbf{S}^A)_{ij} (\mathbf{P}^s\mathbf{S}^B)_{ji} + \frac{1}{2}\sum_{i,j,k,l} \left[\Lambda_{ijkl} - \Lambda_{ijlk}\right] S_{ki}^A S_{lj}^B . \tag{9.72}$$

Here it is taken into account that the first sum in Eq. (7.54) is just Δ_A according to (7.53). The atomic spin population p_A^s in the 3D case obviously is

$$p_A^s = \int w_A(\vec{r}) \varrho_s(\vec{r}) dv = \sum_i (\mathbf{P}^s\mathbf{S}^A)_{ii} . \tag{9.73}$$

*In Ref. 135 a detailed derivation directly in terms of 3D analysis has been given.

Table 9.5
Atomic $\langle \hat{S}^2 \rangle$ components in different molecules obtained from CISD/6-31G** calculations by the fuzzy atoms formalism and "iterative Hirshfeld" weigh functions

Molecule	Atom	$\langle \hat{S}^2 \rangle_A$	Molecule	Atom	$\langle \hat{S}^2 \rangle_A$
H_2	H	0.029	CH_4	C	0.071
Li_2	Li	0.141		H	0.02
Be_2	Be	0.150	C_2H_6	C	0.051
HF	H	0.018		H	0.024
	F	0.019	C_2H_4	C	0.077
H_2O	H	0.021		H	0.026
	O	0.034	C_2H_2	C	0.090
NH_3	N	0.051		H	0.026
	H	0.023			

*[135] permission from American Chemical Society.

Table 9.5 contains the local spin-values computed for simple singlet molecules [135]. It can be seen that in all "usual" molecules the atomic spins are negligible—they represent only minor correlation fluctuations—as one expects on the basis of physical intuition.

A

APPENDICES

A.1 "Mixed" second quantization for non-orthogonal basis functions

A.1.1 Second quantization for orthogonal functions

This appendix cannot provide a complete description of the standard second quantized formalism—for easily accessible introductions for chemists we can recommend [83,136]—so here we can give only a short summary of the main points we will need in the next paragraphs in which we describe the "mixed" version of second quantization utilized in the book. This summary is not intended to be either exhaustive or too strict. Nevertheless, we hope that it might be sufficient to catch the main ideas to the extent necessary for reading the respective parts of the present book, even for the reader who was not familiar with second quantization previously.

Assuming that we are dealing with a set of *m orthonormalized* one-electron spin-orbitals* $\psi_1, \psi_2, \ldots, \psi_m$, then any N-electron wave function which can be written down by using these one-electron functions, represents [5] a linear combination of Slater determinants

$$\Psi_I(1,2,...,N) = \frac{1}{\sqrt{N!}} \begin{vmatrix} \psi_{I_1}(1) & \psi_{I_2}(1) & \psi_{I_3}(1) & \ldots & \psi_{I_N}(1) \\ \psi_{I_1}(2) & \psi_{I_2}(2) & \psi_{I_3}(2) & \ldots & \psi_{I_N}(2) \\ \vdots & \vdots & \vdots & & \vdots \\ \psi_{I_1}(N) & \psi_{I_2}(N) & \psi_{I_3}(N) & \ldots & \psi_{I_N}(N) \end{vmatrix}, \tag{A.1}$$

where the subscript I corresponds to the I-th possible selection $\psi_{I_1}, \psi_{I_2}, \ldots, \psi_{I_N}$ of N spin-orbitals taken out of the whole set of m basis

*If $m \to \infty$, and the functions ψ_μ are properly chosen, then they may represent a complete orthonormalized set. This aspect is, however, not especially important for the present purpose.

spin-orbitals. Here $1, 2, ..., N$ serve as shorthands for the spatial and spin-coordinates of electron 1, 2, and so on, that is $i = \{\vec{r}_i, \sigma_i\}$, where $\vec{r}_i$ is the radius vector of the i-th electron and σ_i is its spin variable.

An important property of any wave function of type (A.1) is that it is *anti-symmetric*, i.e., changes its sign if one interchanges the coordinates of a pair of electrons. Such an interchange of coordinates means interchange of the rows of the determinant on the right-hand side of (A.1), and—according to the known properties of determinants—this causes the change of the sign of the function. Furthermore, the determinant also changes its sign if one interchanges not its rows, but its columns, which is equivalent to changing the order of the spin-orbitals in (A.1). This antisymmetry is of utmost importance because it excludes the possibility that the same spin-orbital could be occupied by two electrons: a determinant with two identical columns vanishes. (The same *spatial* orbital may be filled at most by two electrons, and then the latter must be of opposite spin. This is the classical formulation of the Pauli principle.)

It follows from the above discussion that an antisymmetric wave function of this type is defined by defining the set of the spin-orbitals $\{\psi_{I_\mu}\}$ filled in it, as well as by specifying the *order* of the latter. In the conventional "first quantized" framework this is achieved by using the so-called antisymmetrizing operator $\hat{\mathscr{A}}$ [5] that produces a Slater-determinant (A.1) from the product of the function $\psi_{I_1}(1)\psi_{I_2}(2)\ldots\psi_{I_N}(N)$ that occur along the main diagonal of (A.1). In the "second quantized" formalism (it may also be called "occupation number representation") the same goal is reached in a different manner; that leads to a framework in which many manipulations become easier and more transparent.

The specificity and essential advantage of the second quantized formalism is the fact that the antisymmetry is "built in" at the basic level, and one should not worry about it during the manipulations. The product of functions is commutative, so we need to use some products ("strings") of *operators* that do not commute with each other in order to ensure the antisymmetry. (Their commutation properties will be discussed below.) However, that product of operators must act on some target in order to get an object which can be put into correspondence with a wave function. This specific target is the "vacuum state" containing no electrons; by using the "bra-ket" formalism it can be written as a "ket" $|vac\rangle$; often simply $|0\rangle$, or even $|\rangle$. This vacuum state is an abstract mathematical entity [83]; it should clearly be distinguished from "nothing" (or "zero"), because by definition it is normalized to unity*: $\langle vac|vac\rangle = 1$.

Now, we introduce the *creation operators** $\hat{\psi}_\mu^+$, $\mu = 1,2,\ldots,m$. Applying one $\hat{\psi}_\mu^+$ to the vacuum state, one simply gets a one-electron function

$$\hat{\psi}_\mu^+|vac\rangle \leftrightarrow |\psi_\mu\rangle \,. \tag{A.2}$$

One says that operator $\hat{\psi}_\mu^+$ "creates an electron" in the state ψ_μ; hence the name "creation operator".† Applying two creation operators one gets an *antisymmetric* two-electron wave function

$$\hat{\psi}_\mu^+\hat{\psi}_\nu^+|vac\rangle \leftrightarrow |\psi_\mu\psi_\nu\rangle \,, \tag{A.3}$$

Antisymmetry of this wave function means that if one interchanges the order of operators $\hat{\psi}_\mu^+$ and $\hat{\psi}_\nu^+$ creating electrons in the spin-orbitals ψ_μ and ψ_ν, respectively, then the resulting function changes its sign:

$$\hat{\psi}_\nu^+\hat{\psi}_\mu^+|vac\rangle = -\hat{\psi}_\mu^+\hat{\psi}_\nu^+|vac\rangle \,. \tag{A.4}$$

This shows that the creation operators themselves should *anticommute*, i.e.,

$$\hat{\psi}_\nu^+\hat{\psi}_\mu^+ = -\hat{\psi}_\mu^+\hat{\psi}_\nu^+ \,. \tag{A.5}$$

This can be formulated also in the form that the *anticommutator* of the creation operators vanishes:

$$\{\hat{\psi}_\mu^+, \hat{\psi}_\nu^+\} \equiv \hat{\psi}_\mu^+\hat{\psi}_\nu^+ + \hat{\psi}_\nu^+\hat{\psi}_\mu^+ = 0 \,. \tag{A.6}$$

---

*One can pictorially understand this on the basis of the following analogy. Let us consider a system of some number N of electrons, which are distributed between two subsystems which are infinitely far apart from each other, and it can vary, how the number N is distributed between the two subsystems. If subsystem A contains k electrons, then subsystem B has $N-k$ electrons. The normalization integral of the global system's wave function is—owing to the infinite distance between the subsystems [5]—the product of the normalization integrals "here" and "there": $\langle\Psi|\Psi\rangle = \langle\Psi_k^A|\Psi_k^A\rangle\langle\Psi_{N-k}^B|\Psi_{N-k}^B\rangle$. If we require this general formula to be applicable also for the special case $k = 0$, we should require $\langle\Psi_0^A|\Psi_0^A\rangle = 1$, i.e., define the vacuum state normalized to unity.

*The most common notation for creation operators is $a_\mu^\dagger$. However, we follow Longuet-Higgins [136] to indicate explicitly the set of basis spin-orbitals considered. This may be very convenient when one deals with several sets of basis orbitals simultaneously. (Note that we use the superscript "+" only to denote creation operators; it is to be distinguished from "†" used to denote adjoints.)

†We use the notation "↔" because, strictly speaking, Equations (A.2), (A.3) are not true equalities as the two sides correspond to different representations of the same wave function, but this distinction is irrelevant in our case.

An N-electron antisymmetric wave function of type (A.1), containing the spin-orbitals of the I-th selection $\{\psi_{I_\nu}\}$ of spin-orbitals is obtained similarly, by using N creation operators:

$$|\Psi_I\rangle = \hat{\psi}^+_{I_1}\hat{\psi}^+_{I_2}\ldots\hat{\psi}^+_{I_N}|vac\rangle\ . \tag{A.7}$$

It can easily be seen that the fulfillment of the anticommutation rule (A.5)—or of rule (A.6)—is sufficient to provide antisymmetry (change of the sign) with respect to interchange of any pairs of spin-orbitals (creation operators) in this wave function. In fact, let us interchange the operators which are at the i-th and j-th position in (A.7), and let $j-i=k>0$. This means that we can move the operator $\hat{\psi}^+_{I_i}$ immediately before the operator $\hat{\psi}^+_{I_j}$ by applying the rule (A.5) $k-1$ times to neighboring creation operators. In each such interchange the overall sign is changing. Then we apply the same rule to $\hat{\psi}^+_{I_i}\hat{\psi}^+_{I_j}$, and move $\hat{\psi}^+_{I_j}$ to the original place of $\hat{\psi}^+_{I_i}$ again applying rule (A.5) $k-1$ times to neighboring operators. Altogether we apply rule (A.5) and change the sign $2(k-1)+1$ times, which is an odd number, so to the end we are left with a change of the sign, as a consequence of interchanging two spin-orbitals. It follows immediately from this result that any string of creation operators containing the same operator twice is equal to zero (not the vacuum state, but the "true" zero): a mathematical object that is equal to itself with an opposite sign must vanish.

The "bra" vector, corresponding to the "ket" in the expression (A.7) is, according to the general rules,* the adjoint

$$\langle\Psi_I| = \left(\hat{\psi}^+_{I_1}\hat{\psi}^+_{I_2}\ldots\hat{\psi}^+_{I_N}|vac\rangle\right)^\dagger = \langle vac|(\hat{\psi}^+_{I_N})^\dagger\ldots(\hat{\psi}^+_{I_2})^\dagger(\hat{\psi}^+_{I_1})^\dagger\ . \tag{A.8}$$

Now, we should clarify the meaning of the adjoint operators $(\hat{\psi}^+_\mu)^\dagger$. Obviously, they create electrons when *acting to the left* on the "bra" $\langle vac|$, but we need to find out how they act *to the right*. For that reason, we should consider the overlap integral $\langle\Psi_I|\Psi_J\rangle$ of two determinant wave functions of type (A.1), corresponding to two selections I and J of the spin-orbitals. We know [5,90] that in the most general case this overlap integral is equal to the *determinant* D of the overlap integrals $\langle\psi_{I_\mu}|\psi_{J_\nu}\rangle$ of the one-electron spin-orbitals ψ_{I_μ} and ψ_{J_ν} occupied in the determinants $|\Psi_I\rangle$ and $|\Psi_J\rangle$, respectively:*

*The adjoint of the "ket" is the respective "bra," and the order of operators should be inverted when forming the adjoint: $(\hat{A}\hat{B})^\dagger = \hat{B}^\dagger\hat{A}^\dagger$, etc.

*As the set of spin-orbitals $\{\psi_\mu\}$ is considered orthonormalized, in our case the overlap integral $\langle\Psi_I|\Psi_J\rangle$ is equal ±1 if the sets $\{\psi_{I_\mu}\}$ and $\{\psi_{J_\nu}\}$ are the same but may differ in the order in which the individual spin-orbitals occur, or it is equal to 0 if these sets differ at least by one spin-orbital.

$$\langle\Psi_I|\Psi_J\rangle = D = Det\{\langle\psi_{I\mu}|\psi_{J\nu}\rangle\} \; . \tag{A.9}$$

We note that in this appendix all expressions refer to spin-orbitals and, accordingly, all the integrals include summation over the spin variables.

To get an insight into the properties of $(\hat{\psi}_\mu^+)^\dagger$, let us first consider the specific case when the first spin-orbitals in the two sets are the same: $\psi_{I_1} = \psi_{J_1}$. This means that the diagonal $(1,1)$ element of the determinant D is 1, and all the other elements in the first row and first column are 0, as the spin-orbitals used are orthonormalized and neither functions $|\Psi_I\rangle$ and $|\Psi_J\rangle$ can contain the same spin-orbital twice. Using the standard technique of manipulations with determinants, we can now *expand* determinant D along its first row (or column). As the first row (column) contains only a single non-zero term in the $(1,1)$ position and it is equal 1, we get

$$\begin{aligned}\langle\Psi_I|\Psi_J\rangle &= \langle vac|(\hat{\psi}_{I_N}^+)^\dagger \dots (\hat{\psi}_{I_2}^+)^\dagger (\hat{\psi}_{I_1}^+)^\dagger \hat{\psi}_{I_1}^+ \hat{\psi}_{J_2}^+ \dots \hat{\psi}_{J_N}^+|vac\rangle \\ &= D = 1 \cdot M(1|1) \; , \end{aligned} \tag{A.10}$$

where $M(1|1)$ is the minor (determinant of order $N-1$) obtained by crossing out the first row and first column of the determinant D.

It is easy to see, that the minor $M(1|1)$ is the determinant which one obtains by calculating the overlap of the wave functions containing one electron less (spin-orbitals ψ_{I_1} and ψ_{J_1} are absent):

$$M(1|1) = \langle vac|(\hat{\psi}_{I_N}^+)^\dagger \dots (\hat{\psi}_{I_2}^+)^\dagger \hat{\psi}_{J_2}^+ \dots \hat{\psi}_{J_N}^+|vac\rangle \; . \tag{A.11}$$

This means that the adjoint operator $(\hat{\psi}_{I_1}^+)^\dagger$ effectively "crosses out" operator $\hat{\psi}_{I_1}^+$, or as it is usual to say, annihilates the electron which was created by the latter. Accordingly, it is called "annihilation operator," and one denotes accordingly*

$$\hat{\psi}_\mu^- = (\hat{\psi}_\mu^+)^\dagger \; . \tag{A.12}$$

Thus the "bra" function in Eq. (A.8) can also be written as

$$\langle\Psi_I| = \langle vac|\hat{\psi}_{I_N}^- \dots \hat{\psi}_{I_2}^- \hat{\psi}_{I_1}^- \; . \tag{A.13}$$

However, the annihilation operator is not simply the inverse of the respective creation operator. Instead, one should apply the following set of rules. First one has

$$\{\hat{\psi}_\mu^-, \hat{\psi}_\nu^-\} \equiv \hat{\psi}_\mu^- \hat{\psi}_\nu^- + \hat{\psi}_\nu^- \hat{\psi}_\mu^- = 0 \; , \tag{A.14}$$

*The more usual notation is simply a_μ, but we, as noted above, prefer to use the more detailed notations of Longuet-Higgins.

which follows from the definition (A.12) and the formula (A.6) for anticommutator of the creation operators. Then we have the so-called "Fermion anti-commutation rule" for creation and an annihilation operators:

$$\{\hat{\psi}_\mu^+, \hat{\psi}_\nu^-\} \equiv \hat{\psi}_\mu^+ \hat{\psi}_\nu^- + \hat{\psi}_\nu^- \hat{\psi}_\mu^+ = \delta_{\mu\nu} . \qquad (A.15)$$

Additional rules are

$$\hat{\psi}_\mu^- |vac\rangle = 0 ; \quad \text{and} \quad \langle vac| \hat{\psi}_\mu^+ = 0 , \qquad (A.16)$$

which express that one cannot annihilate an electron from the vacuum. (In accord with Eqs. (A.8) and (A.12), the role of creation and annihilation operators is interchanged when applied to the left, so $\hat{\psi}_\mu^-$ creates and $\hat{\psi}_\mu^+$ annihilates an electron when applied to a "bra.")

It is easy to see that these rules form a consistent set. The anticommutation $\hat{\psi}_\mu^+ \hat{\psi}_\nu^- = -\hat{\psi}_\nu^- \hat{\psi}_\mu^+$, following from Eq. (A.15) if $\mu \neq \nu$, means that in the case when the respective creation operator $\hat{\psi}_\nu^+$ is not the first in the string of creation operators considered, then the annihilation operator $\hat{\psi}_\nu^-$ can be moved to the right to meet it with as many changes of sign as would appear if one moved $\hat{\psi}_\nu^+$ to the left to be placed immediately to the right of $\hat{\psi}_\nu^-$.

The anticommutation rule (A.15) can also be written as

$$\hat{\psi}_\nu^- \hat{\psi}_\mu^+ = \delta_{\mu\nu} - \hat{\psi}_\mu^+ \hat{\psi}_\nu^- . \qquad (A.17)$$

This means that when acting with $\hat{\psi}_\nu^-$ on a string containing $\hat{\psi}_\nu^+$, then one gets that string, but with $\hat{\psi}_\nu^+$ omitted and multiplied by a factor ± 1 (the sign depends on the number of interchanges necessary to reach $\hat{\psi}_\nu^+$), as well as a term with $\hat{\psi}_\nu^-$ placed to the right with respect to $\hat{\psi}_\nu^+$. After the respective number of interchanges, operator $\hat{\psi}_\nu^-$ will reach the "ket" $|vac\rangle$, and be eliminated in accord with the first rule (A.16)—it cannot meet another $\hat{\psi}_\nu^+$, because, as noted above, no strings may contain the same creation operator more than once.

It follows from these rules, that the overlap integral

$$\langle \Psi_I | \Psi_J \rangle = \langle vac| \hat{\psi}_{I_N}^- \dots \hat{\psi}_{I_2}^- \hat{\psi}_{I_1}^- \hat{\psi}_{J_1}^+ \hat{\psi}_{J_2}^+ \dots \hat{\psi}_{J_N}^+ |vac\rangle \qquad (A.18)$$

can differ from zero only if for every annihilation operator $\hat{\psi}_{I_\mu}^-$ in the "bra" one can find a respective creation operator $\hat{\psi}_{I_\mu}^+$ in the "ket," i.e., the sets of spin-orbitals $\{\psi_{I_\mu}\}$ may differ from the set $\{\psi_{J_\nu}\}$ at most by the order of the terms.*

*This is in full accord with the fact that the determinant in Eq. (A.9) would vanish if the set $\{\psi_{I_\mu}\}$ contained one or more spin-orbitals orthogonal to all spin-orbitals in the set $\{\psi_{J_\nu}\}$, as that would involve one or more rows in the determinant (A.9) containing only zeros.

It is easy to see that acting with the product $\hat{\psi}^+_\mu \hat{\psi}^-_\mu$ on a wave function of type $\hat{\psi}^+_{J_1} \hat{\psi}^+_{J_2} \dots \hat{\psi}^+_{J_N} |vac\rangle$ will reproduce it unchanged if the string of creation operators does contain $\hat{\psi}^+_\mu$: operator $\hat{\psi}^-_\mu$ creates a string with $\hat{\psi}^+_\mu$ "crossed out" and with a sign depending on the position of the latter in the string; then $\hat{\psi}^+_\mu$ may be put back to the original place with the same number of interchanges, so the final sign will be positive. If, however, the string does not contain $\hat{\psi}^+_\mu$, then the result of acting by $\hat{\psi}^+_\mu \hat{\psi}^-_\mu$ will be zero. (One cannot annihilate an electron in a state in which it was not created beforehand.)

Now, we can form the sum

$$\sum_\mu \hat{\psi}^+_\mu \hat{\psi}^-_\mu \tag{A.19}$$

where the sum runs over all the spin-orbitals of the (finite or infinite) orthonormalized spin-orbital basis $\{\psi_\mu\}$. When acting on the wave function of type $\hat{\psi}^+_{I_1} \hat{\psi}^+_{I_2} \dots \hat{\psi}^+_{I_N} |vac\rangle$ each term of the sum will produce a factor 0 or 1 depending on whether or not $\hat{\psi}^+_\mu$ is among the set $\{\psi^+_{I_\nu}\}$. As the sum is over the whole basis, we shall get as many terms with factor 1 as the number of electrons N in the wave function considered. This means that every N-electron wave function will be an eigenfunction of this sum of operators with an eigenvalue N; therefore one calls it the "operator of number of electrons" $\hat{N}$:

$$\hat{N} = \sum_\mu \hat{\psi}^+_\mu \hat{\psi}^-_\mu \, . \tag{A.20}$$

An important advantage of the second quantized formalism is that we need not specify beforehand the number of the electrons, and can obtain expressions that are valid in the general case of any number of electrons. Operator $\hat{N}$ gives a specific example of that, but it is also true for other operators, including the Hamiltonian.

To see that, let us first consider a one-electron operator $\hat{h}$ by using "bra-ket" notations. Applying twice the "resolution of identity"

$$\hat{I} = \sum_\mu |\psi_\mu\rangle\langle\psi_\mu| \tag{A.21}$$

where μ again runs over all the spin-orbitals of the basis considered, operator $\hat{h}$ can be presented as*

$$\hat{h} = \hat{I}\hat{h}\hat{I} = \sum_{\mu,\nu} |\psi_\mu\rangle\langle\psi_\mu|\hat{h}|\psi_\nu\rangle\langle\psi_\nu| = \sum_{\mu,\nu} h_{\mu\nu}|\psi_\mu\rangle\langle\psi_\nu| \, , \tag{A.22}$$

*Strictly speaking, if the basis is finite (not complete), then this means replacing operator $\hat{h}$ by its "projection to the basis $\{\psi_\mu\}$." From our point of view that distinction has no practical significance.

where $h_{\mu\nu} = \langle\psi_\mu|\hat{h}|\psi_\nu\rangle$, is the matrix element of operator $\hat{h}$ between spin-orbitals ψ_μ, ψ_ν, and operator $|\psi_\mu\rangle\langle\psi_\nu|$ replaces electron on spin-orbital ψ_ν by an electron on spin-orbital ψ_μ; it may be called a "shift-operator."

Now, it is easy to see that the second quantized analogue of that shift operator is the operator $\hat{\psi}_\mu^+\hat{\psi}_\nu^-$. However, contrary to the operator $|\psi_\mu\rangle\langle\psi_\nu|$, it can be applied not only to a one-electron wave function but also to an N-electron one, formed by a string of N creation operators acting on the vacuum state $|vac\rangle$. Then the one-electron part of the Hamiltonian, in the form applicable independently of the actual number of electrons can be written in a second quantized framework as

$$\hat{H}_1 = \sum_{\mu,\nu} h_{\mu\nu}\hat{\psi}_\mu^+\hat{\psi}_\nu^- \, . \tag{A.23}$$

It is easy to see that the matrix elements of this operator between two determinant wave functions reproduce the known Slater rules [5] valid for determinant wave functions built up of orthonormalized sets of spin-orbitals. For that reason we should evaluate the matrix elements $\langle\Psi_I|\hat{H}_1|\Psi_J\rangle$. Here $\langle\Psi_I|$ is given by Eq. (A.13), while $\hat{H}_1|\Psi_J\rangle$ is equal to

$$\hat{H}_1|\Psi_J\rangle = \sum_{\mu,\nu} h_{\mu\nu}\hat{\psi}_\mu^+\hat{\psi}_\nu^-\hat{\psi}_{J_1}^+\hat{\psi}_{J_2}^+\cdots\hat{\psi}_{J_N}^+|vac\rangle \, . \tag{A.24}$$

Similar to the case of $\hat{\psi}_\mu^+\hat{\psi}_\mu^-$ discussed above, operator $\hat{\psi}_\mu^+\hat{\psi}_\nu^-$ gives a non-vanishing result only if it acts on a string of creation operators that contains $\hat{\psi}_\nu^+$; in that case the latter is replaced by $\hat{\psi}_\mu^+$. This means that the "ket" in the right-hand side contains strings that are identical to that in $|\Psi_J\rangle$ (in the cases when $\mu = \nu$) or differs from it by one replacement. As the overlap integrals of type (A.18) can differ from zero only in the case when for every annihilation operator in the "bra" there is a respective creation operator in the "ket," we see that the matrix element $\langle\Psi_I|\hat{H}_1|\Psi_J\rangle$ will be non-zero only if the sets $\{\psi_{I\nu}\}$ and $\{\psi_{J\nu}\}$ are either the same or if they differ by only one spin-orbital.* In the former case, i.e., if we have the same function in the "bra" and in the "ket" (all $I_\nu = J_\nu$), for every $\psi_{I\nu}^+$ in $|\Psi_I\rangle$ there occurs a term $h_{\mu\mu}\hat{\psi}_\mu^+\hat{\psi}_\mu^-$ with $\mu = I_\nu$, as the sum for μ runs over the entire basis, so the resulting matrix element will be

$$\langle\Psi_I|\hat{H}_1|\Psi_I\rangle = \sum_{\nu=1}^{N} h_{I_\nu I_\nu} \, . \tag{A.25}$$

*We assume that the creation operators have been ordered so as to achieve the maximum coincidence. That generates a factor of ± 1 only.

If the two functions differ by one spin-orbital—some $\hat{\psi}^+_{I\nu}$ being replaced by a $\hat{\psi}^+_{J\nu}$—then the only term in the operator (A.23) will lead to non-zero effect that restores $\hat{\psi}^+_{I\nu}$ in the place of $\hat{\psi}^+_{J\nu}$; then the resulting matrix element will obviously equal $h_{I_\nu J_\nu}$.

The two-electron part of the Hamiltonian can analogously be written down in terms of the *two-electron integrals:*

$$\hat{H}_2 = \frac{1}{2} \sum_{\mu,\nu,\rho,\tau} [\psi_\mu \psi_\nu | \psi_\rho \psi_\tau] \hat{\psi}^+_\mu \hat{\psi}^+_\nu \hat{\psi}^-_\tau \hat{\psi}^-_\rho \,, \tag{A.26}$$

where the two-electron integral is in the $[12|12]$ convention and the integrations again include summation over the spin variables:

$$[\psi_\mu \psi_\nu | \psi_\rho \psi_\tau] = \int\int \psi^*_\mu(1) \psi^*_\nu(2) \frac{1}{r_{12}} \psi_\rho(1) \psi_\tau(2) d\tau_1 d\tau_2 \,, \tag{A.27}$$

with the "volume elements" $d\tau_i = dv_i d\sigma_i$. Note the change of the order of the subscripts in the operator string as compared to that in the two-electron integral.

As there are two creation and two annihilation operators in the two-electron part of the Hamiltonian Eq. (A.26), it can have non-zero matrix elements between determinants differing by not more than two spin-orbitals. Considering these cases, one recovers again the known Slater rules. We shall not go through them here in detail, only briefly discuss the appearance of the so-called "exchange terms." As all the summation indices in Eq. (A.26) run over the whole spin-orbital basis, the result of acting with $\hat{H}_2$ will always contain pairs of terms that differ only in the interchange of subscripts ρ and τ. Owing to the antisymmetry (A.5), the respective wave functions differ only in sign—but the respective integrals are, in general, quite different. The terms for which one has to perform the interchange of the creation operators in the "ket" in order to put them into full correspondence with the annihilation operators in the "bra," are called "exchange" ones; they carry a negative sign in the expressions of the matrix elements $\langle \Psi_I | \hat{H}_2 | \Psi_J \rangle$.*

As is known, the expectation values of $\hat{H}_1$ and $\hat{H}_2$ can be expressed with the help of the matrix-representations **P** and **Γ** of the first- and second-order density matrices, respectively:

$$\langle \hat{H}_1 \rangle = Tr(\mathbf{hP}) = \sum_{\mu,\nu} h_{\mu\nu} P_{\nu\mu} \,; \tag{A.28}$$

*As the summation over the spin variables in the integral (A.27) gives non-zero value for the integrals entering both terms containing a given pair of interchanged creation operators only if all the spin parts of the four spin-orbitals are the same (either α or β), the exchange phenomenon appears only for electrons of the same spin.

and

$$\langle \hat{H}_2 \rangle = \frac{1}{2} \sum_{\mu,\nu,\tau\rho} [\psi_\mu \psi_\nu | \psi_\rho \psi_\tau] \Gamma_{\rho\tau\mu\nu} \; . \tag{A.29}$$

Comparing with Eqs. (A.23) and (A.26) we see that the matrix elements of the first- and second-order density matrix can be expressed as the expectation values of the operator strings

$$P_{\nu\mu} = \langle \hat{\psi}_\mu^+ \hat{\psi}_\nu^- \rangle \; ; \tag{A.30}$$

and

$$\Gamma_{\rho\tau\mu\nu} = \langle \hat{\psi}_\mu^+ \hat{\psi}_\nu^+ \hat{\psi}_\tau^- \hat{\psi}_\rho^- \rangle \; , \tag{A.31}$$

calculated for the actual wave function. (Note that $\Gamma_{\rho\tau\mu\nu} = \Gamma_{\tau\rho\nu\mu} = -\Gamma_{\tau\rho\mu\nu}$, etc.)

A.1.2 "Mixed" second quantization for non-orthogonal functions

In quantum chemical practice we are mostly working with atom-centered basis functions that are not orthogonal. If the basis is reasonably selected then the assignments to the basis functions to the individual atoms and the interatomic overlaps may carry useful information of a chemical nature, so it is worth working explicitly with that non-orthogonal atomic basis. The second quantized formalism outlined briefly in the previous paragraph needs some modifications in order to work directly with overlapping basis functions. That modified formalism can be developed on the basis of the standard (orthogonal) one, and its characteristic feature is the distinction between the adjoint of a creation operator and the "effective" annihilation operator that behaves like a conventional annihilation operator does in the orthogonal case—in the sense of obeying the Fermion anticommutation rule crucial for the formalism to be practically usable.

In the following we shall assume that the basis of atomic spin-orbitals $\{\chi_\mu\}$ is overlapping but not redundant (not linearly dependent). Then we can define in a standard manner also an orthonormalized basis $\{\psi_\mu\}$ that spans *the same (sub)space* of one-electron functions. The most adequate choice is the Löwdin-orthogonalized set of functions

$$\psi_\mu = \sum_\nu (\mathbf{S}^{-\frac{1}{2}})_{\nu\mu} \chi_\nu \; , \tag{A.32}$$

where $\mathbf{S}$ is the overlap matrix of the basis spin-orbitals having the elements $S_{\mu\nu} = \langle \chi_\mu | \chi_\nu \rangle$. The original basis functions, in turn, can be expressed in terms of the orthogonalized ones as

$$\chi_\mu = \sum_\nu (\mathbf{S}^{\frac{1}{2}})_{\nu\mu} \psi_\nu \ . \tag{A.33}$$

We can introduce in standard manner the creation and annihilation operators $\hat{\psi}_\mu^+$ and $\hat{\psi}_\mu^-$ for the orthonormalized set $\{\psi_\mu\}$. The creation operator $\hat{\chi}_\mu^+$ should create an electron on the basis spin-orbital χ_μ; in accord with Eq. (A.33) it should be

$$\hat{\chi}_\mu^+ = \sum_\nu (\mathbf{S}^{\frac{1}{2}})_{\nu\mu} \hat{\psi}_\nu^+ \ . \tag{A.34}$$

As a consequence of the anticommutation relationships (A.5), (A.6) these operators also anticommute.

A *formal* annihilation operator $\hat{\chi}_\mu^-$ can be defined as the adjoint of $\hat{\chi}_\mu^+$:

$$\hat{\chi}_\mu^- = (\hat{\chi}_\mu^+)^\dagger = \sum_\nu (\mathbf{S}^{\frac{1}{2}})_{\mu\nu} \hat{\psi}_\nu^- \ . \tag{A.35}$$

These operators also anticommute. However, they do not obey the Fermion anticommutation rule; instead. one has

$$\begin{aligned}
\{\hat{\chi}_\mu^+, \hat{\chi}_\nu^-\} &= \hat{\chi}_\mu^+ \hat{\chi}_\nu^- + \hat{\chi}_\nu^- \hat{\chi}_\mu^+ \\
&= \sum_\rho (\mathbf{S}^{\frac{1}{2}})_{\rho\mu} \hat{\psi}_\rho^+ \sum_\tau (\mathbf{S}^{\frac{1}{2}})_{\nu\tau} \hat{\psi}_\tau^- + \sum_\iota (\mathbf{S}^{\frac{1}{2}})_{\nu\tau} \hat{\psi}_\tau^- \sum_\rho (\mathbf{S}^{\frac{1}{2}})_{\rho\mu} \hat{\psi}_\rho^+ \\
&= \sum_{\rho,\tau} (\mathbf{S}^{\frac{1}{2}})_{\nu\tau} \{\hat{\psi}_\rho^+, \hat{\psi}_\tau^-\} (\mathbf{S}^{\frac{1}{2}})_{\rho\mu} = \sum_{\rho,\tau} (\mathbf{S}^{\frac{1}{2}})_{\nu\tau} \delta_{\tau\rho} (\mathbf{S}^{\frac{1}{2}})_{\rho\mu} = S_{\nu\mu} \ .
\end{aligned} \tag{A.36}$$

In such circumstances it is worth introducing the *biorthogonal* set of spin-orbitals $\{\varphi_\mu\}^*$

$$\varphi_\mu = \sum_\nu (\mathbf{S}^{-1})_{\nu\mu} \chi_\nu = \sum_\rho (\mathbf{S}^{-\frac{1}{2}})_{\rho\mu} \psi_\rho \ , \tag{A.37}$$

that span the same (sub)space as the spin-orbitals $\{\chi_\nu\}$ and $\{\psi_\rho\}$, and the respective creation operators

$$\hat{\varphi}_\mu^+ = \sum_\nu (\mathbf{S}^{-1})_{\nu\mu} \hat{\chi}_\nu^+ = \sum_\rho (\mathbf{S}^{-\frac{1}{2}})_{\rho\mu} \hat{\psi}_\rho^+ \ . \tag{A.38}$$

Then the adjoints of the creation operators

$$\hat{\varphi}_\mu^- = (\hat{\varphi}_\mu^+)^\dagger = \sum_\nu (\mathbf{S}^{-1})_{\mu\nu} \hat{\chi}_\nu^- = \sum_\rho (\mathbf{S}^{-\frac{1}{2}})_{\mu\rho} \hat{\psi}_\rho^- \ , \tag{A.39}$$

*One often uses also the notation $\widetilde{\chi}_\mu$ instead of φ_μ, e.g., that is adapted in [83].

will behave as the "effective annihilation operators" for spin-orbitals χ_ν, because, as is easy to see, they permit to recover the Fermion anticommutation rules as

$$\{\hat{\chi}_\mu^+, \hat{\varphi}_\nu^-\} \equiv \hat{\chi}_\mu^+ \hat{\varphi}_\nu^- + \hat{\varphi}_\nu^- \hat{\chi}_\mu^+ = \delta_{\mu\nu} . \tag{A.40}$$

This means that *the annihilation operator $\hat{\varphi}_\mu^-$ defined with respect to the biorthogonal set of spin-orbitals is the operator that acts in the non-orthogonal case exactly in the same manner as the usual annihilation operators do in the orthogonal one.* As a consequence, the "shift operator" replacing the spin-orbital χ_μ by the spin-orbital χ_ν is the "mixed" expression $\hat{\chi}_\nu^+ \hat{\varphi}_\mu^-$ if an overlapping basis is used. (This property is generalized when writing Eq. (7.45) in the analysis of local spins.)

The "operator of the number of electrons," $\hat{N}$, can be obtained in the "mixed" formalism by starting from Eq. (A.20). One expresses the creation operators $\hat{\psi}_\mu^+$ through the operators $\hat{\chi}_\rho^+$ by inverting Eq. (A.34) as

$$\hat{\psi}_\mu^+ = \sum_\rho (\mathbf{S}^{-\frac{1}{2}})_{\rho\mu} \hat{\chi}_\rho^+ , \tag{A.41}$$

and operators $\hat{\psi}_\mu^-$ through the operators $\hat{\varphi}_\nu^-$ by inverting Eq. (A.39) as

$$\hat{\psi}_\mu^- = \sum_\nu (\mathbf{S}^{\frac{1}{2}})_{\mu\nu} \hat{\varphi}_\nu^- , \tag{A.42}$$

Substituting Eqs. (A.41) and (A.42) into Eq. (A.20), one trivially obtains

$$\hat{N} = \sum_\nu \hat{\chi}_\nu^+ \hat{\varphi}_\nu^- . \tag{A.43}$$

When doing Hilbert-space analysis, each basis orbital is assigned to one of the atoms. Then Eq. (A.43) can be rewritten as

$$\hat{N} = \sum_A \hat{N}_A , \tag{A.44}$$

where

$$\hat{N}_A = \sum_{\nu \in A} \hat{\chi}_\nu^+ \hat{\varphi}_\nu^- , \tag{A.45}$$

is the *operator of the atomic population* for atom A [24].

The expectation value of operator $\hat{N}_A$ is Mulliken's gross atomic population Q_A of atom A:

$$\langle \hat{N}_A \rangle = Q_A = \sum_{\mu \in A} (\mathbf{PS})_{\mu\mu} . \tag{A.46}$$

where **P** and **S** are the so-called "density matrix" and the overlap matrix for the spin-orbital basis $\{\chi_\mu\}$.* To see that, we determine the general expression for the expectation value $\langle \hat{\chi}_\nu^+ \hat{\varphi}_\mu^- \rangle$, following [26].

For any wave function which can be expanded in terms of the one-electron basis $\{\chi_\mu\}$—whether or not it is a single determinant—the distribution $\varrho(\vec{r},s)$ of the electron charge and spin density in the physical space can be expressed in terms of the basis functions and the elements of the "density matrix" **P**:

$$\varrho(\vec{r},s) = \sum_{\mu,\nu} P_{\nu\mu} \chi_\mu^*(\vec{r},s) \chi_\nu(\vec{r},s) \tag{A.47}$$

On the other hand, $\varrho(\vec{r},s)$ is the expectation value of the operator $\hat{\varrho}(\vec{r},s) = \sum_i \delta(\vec{r}-\vec{r}_i)\delta_{ss_i}$. In the Löwdin-orthogonalized basis $\{\psi_i\}$, the second quantized representation of this operator is quite similar to that of the one-electron part of the Hamiltonian (A.23); just the integrals of the one-electron Hamiltonian $h_{\mu\nu}$ have to be replaced by the integrals of the density operator. Thus

$$\hat{\varrho}(\vec{r},s) = \sum_{\mu,\nu} \langle \psi_\mu(1) | \delta(\vec{r}-\vec{r}_i)\delta_{ss_i} | \psi_\nu(1) \rangle \hat{\psi}_\mu^+ \hat{\psi}_\nu^- \ . \tag{A.48}$$

By substituting Eq. (A.32) for the functions ψ_μ, ψ_ν, and expressing the creation and annihilation operators $\hat{\psi}_\mu^+$ and $\hat{\psi}_\nu^-$ via the operators $\hat{\chi}_\mu^+$ and $\hat{\varphi}_\nu^-$ according to Eqs. (A.34) and (A.39), respectively, one obtains after a simple algebra

$$\begin{aligned}\hat{\varrho}(\vec{r},s) &= \sum_{\mu,\nu,\rho} S_{\rho\mu}^{-1} \langle \chi_\mu(1) | \delta(\vec{r}-\vec{r}_1)\delta_{ss_1} | \chi_\nu(1) \rangle \hat{\chi}_\rho^+ \hat{\varphi}_\nu^- \\ &= \sum_{\mu,\nu,\rho} S_{\rho\mu}^{-1} \chi_\mu^*(\vec{r},s) \chi_\nu(\vec{r},s) \hat{\chi}_\rho^+ \hat{\varphi}_\nu^- \ .\end{aligned} \tag{A.49}$$

Here and further on we use the shorthand $S_{\rho\mu}^{-1} = (\mathbf{S}^{-1})_{\rho\mu}$. By taking the expectation value, one gets

$$\varrho(\vec{r},s) = \sum_{\mu,\nu,\rho} S_{\rho\mu}^{-1} \langle \hat{\chi}_\rho^+ \hat{\varphi}_\nu^- \rangle \chi_\mu^*(\vec{r},s) \chi_\nu(\vec{r},s) \ . \tag{A.50}$$

Comparison with Eq. (A.47) gives

$$P_{\nu\mu} = \sum_\rho S_{\rho\mu}^{-1} \langle \hat{\chi}_\rho^+ \hat{\varphi}_\nu^- \rangle \ , \tag{A.51}$$

---

*As discussed in Section 4.1, matrix **P** is the basis-representation of the first-order density matrix if the basis is orthonormalized, but the situation is somewhat more complicated if it is not: in that case it is matrix **PS** that may be considered to be the proper representation of the first-order density matrix [27]. Nonetheless, one traditionally calls matrix **P** "density matrix" even if the basis is overlapping.

from which, after simple manipulations, one gets

$$\langle \hat{\chi}_\mu^+ \hat{\varphi}_\nu^- \rangle = \sum_\tau P_{\nu\tau} S_{\tau\mu} = (\mathbf{PS})_{\nu\mu} \,, \tag{A.52}$$

Analogously, one has for the second-order density matrix in the overlapping case the expression [83]

$$\langle \hat{\chi}_\mu^+ \hat{\chi}_\nu^+ \hat{\varphi}_\sigma^- \hat{\varphi}_\lambda^- \rangle = \sum_{\eta,\tau} \Gamma_{\lambda\sigma\eta\tau} S_{\eta\mu} S_{\tau\nu} \,. \tag{A.53}$$

If the wave function is a single determinant, this can be expressed as a sum of a "direct" (or "Coulombic") and an "exchange" terms [26, 67]:

$$\langle \hat{\chi}_\mu^+ \hat{\chi}_\nu^+ \hat{\varphi}_\sigma^- \hat{\varphi}_\lambda^- \rangle = (\mathbf{PS})_{\lambda\mu} (\mathbf{PS})_{\sigma\nu} - (\mathbf{PS})_{\lambda\nu} (\mathbf{PS})_{\sigma\mu} \,. \tag{A.54}$$

This expression also leads to ($\mu \neq \nu$ because $A \neq B$):

$$\begin{aligned}\langle \hat{N}_A \hat{N}_B \rangle &= \sum_{\mu\in A} \sum_{\nu\in B} \langle \chi_\mu^+ \varphi_\mu^- \chi_\nu^+ \varphi_\nu^- \rangle = -\sum_{\mu\in A} \sum_{\nu\in B} \langle \chi_\mu^+ \chi_\nu^+ \varphi_\mu^- \varphi_\nu^- \rangle \\ &= -\sum_{\mu\in A} \sum_{\nu\in B} \left[(\mathbf{PS})_{\nu\mu} (\mathbf{PS})_{\mu\nu} - (\mathbf{PS})_{\nu\nu} (\mathbf{PS})_{\mu\mu} \right] \,,\end{aligned} \tag{A.55}$$

from which one obtains Eq. (6.38) by taking into account Eq. (A.46) and turning to the spatial orbitals.

For completeness, we shall also discuss briefly the second quantized Hamiltonian for the overlapping case. There are several equivalent expressions that can be written down; we quote here that one which contains the one- and two-electron integrals calculated for the original overlapping basis, as well as the creation and "effective" annihilation operators discussed above. (That form has been utilized in our "chemical Hamiltonian approach" [26].) We start from expressions (A.23) and (A.26) for the one- and two-electron parts of the Hamiltonian written down for the Löwdin-orthogonalized basis $\{\psi_\mu\}$, express the orbitals in the integrals through those of the overlapping basis $\{\chi_\nu\}$ by using Eq. (A.32), and the creation and annihilation operators $\hat{\psi}_\mu^+$, $\hat{\psi}_\mu^-$ through the operators $\hat{\chi}_\mu^+$ and $\hat{\varphi}_\mu^-$, respectively, by using Eqs.(A.41) and (A.42), and obtain (adding also the nuclear repulsion term)

$$\begin{aligned}\hat{H} = \sum_{A<B} \frac{Z_A Z_B}{R_{AB}} &+ \sum_{\lambda,\varepsilon,\sigma} S_{\sigma\lambda}^{-1} h_{\lambda\varepsilon} \hat{\chi}_\sigma^+ \hat{\varphi}_\varepsilon^- \\ &+ \tfrac{1}{2} \sum_{\gamma,\nu,\kappa,\rho,\eta,\varepsilon} S_{\eta\gamma}^{-1} S_{\varepsilon\nu}^{-1} [\chi_\gamma \chi_\nu | \chi_\kappa \chi_\rho] \hat{\chi}_\eta^+ \hat{\chi}_\varepsilon^+ \hat{\varphi}_\rho^- \hat{\varphi}_\kappa^-\end{aligned} \tag{A.56}$$

By calculating the expectation values of the operator strings in this equation, the overlap matrices and inverse overlap matrices cancel and one obtains for the energy an expression quite resembling the orthogonal case:

$$E = \sum_{A<B} \frac{Z_A Z_B}{R_{AB}} + \sum_{\lambda,\varepsilon} h_{\lambda\varepsilon} P_{\varepsilon\lambda} + \frac{1}{2} \sum_{\gamma,\nu,\kappa,\rho} [\chi_\gamma \chi_\nu | \chi_\kappa \chi_\rho] \Gamma_{\kappa\rho\gamma\nu} \,. \tag{A.57}$$

Finally, it may be noted that using Eq. (A.38) connecting the creation operators $\hat{\varphi}_\mu^+$ and $\hat{\chi}_\nu^+$ through the elements of the inverse overlap matrix $\mathbf{S}^{-1}$, one can rewrite Eqs. (A.52) and (A.53) in a form resembling *formally* Eqs. (A.30) and (A.31), valid in the case of an orthonormal basis set:

$$P_{\nu\mu} = \langle \hat{\varphi}_\mu^+ \hat{\varphi}_\nu^- \rangle \,; \tag{A.58}$$

and

$$\Gamma_{\rho\tau\mu\nu} = \langle \hat{\varphi}_\mu^+ \hat{\varphi}_\nu^+ \hat{\varphi}_\tau^- \hat{\varphi}_\rho^- \rangle \,. \tag{A.59}$$

Analogously, the Hamiltonian (A.56 in terms of operators $\hat{\varphi}_\mu^+$, $\hat{\varphi}_\nu^-$ can be written as

$$\hat{H} = \sum_{A<B} \frac{Z_A Z_B}{R_{AB}} + \sum_{\lambda,\varepsilon} h_{\lambda\varepsilon} \hat{\varphi}_\varepsilon^+ \hat{\varphi}_\lambda^- + \frac{1}{2} \sum_{\gamma,\nu,\kappa,\rho} [\chi_\gamma \chi_\nu | \chi_\kappa \chi_\rho] \hat{\varphi}_\gamma^+ \hat{\varphi}_\nu^+ \hat{\varphi}_\rho^- \hat{\varphi}_\kappa^- \,. \tag{A.60}$$

A.2 Calculation of Becke's weight functions

Becke's $w_A(\vec{r})$ is an algebraic function which strictly satisfies requirement (9.3). In addition, it gives exactly $w_A = 1$ on the "own" nucleus A—all the other w_B-s are zero there—c.f. Eq. (9.7). Becke's [123] formulae were collected in the appendix of [122]. (Reproduced with permission from Elsevier.)

One starts by introducing the definitions

$$\chi_{AB} = r_A^0 / r_B^0 \; ; \tag{A.61}$$

$$a_{AB} = 0.25(1-\chi_{AB}^2)/\chi_{AB}, \quad \text{but } |a_{AB}| \leq 0.5 \; ; \tag{A.62}$$

$$r_A = |\vec{r} - \vec{R}_A| \; ; \qquad r_B = |\vec{r} - \vec{R}_B| \; ; \qquad R_{AB} = |\vec{R}_A - \vec{R}_B| \; ; \tag{A.63}$$

$$\mu(r_A, r_B) = (r_A - r_B)/R_{AB} \; . \tag{A.64}$$

The $w_A(\vec{r})$ is defined as

$$w_A(\vec{r}) = \frac{P_A(\vec{r})}{\sum\limits_B P_B(\vec{r})} \; ; \tag{A.65}$$

with

$$P_A(\vec{r}) = \prod_{\substack{B \\ B \neq A}} 0.5[1 - \nu^{(k)}(r_A, r_B)] \; . \tag{A.66}$$

$\nu^{(k)}(r_A, r_B)$ is calculated iteratively as

$$\nu^{(l)}(r_A, r_B) = \nu^{(l-1)}(r_A, r_B) \left(1.5 - 0.5[\nu^{(l-1)}(r_A, r_B)]^2\right) \; ; \tag{A.67}$$

starting with

$$\nu^{(0)}(r_A, r_B) = \mu(r_A, r_B) + a_{AB}\left(1 - [\mu(r_A, r_B)]^2\right) \; . \tag{A.68}$$

Here k is a fixed parameter of the procedure (number of iterations) determining the stiffness of the cutoff (usually $k = 3$), and the r_A^0s are the fixed atomic radii. ($\vec{R}_A$ are the radius-vectors of the nuclei.)

Bibliography

1. G.G. Hall, *Chairman's remarks,* 5th International Congress on Quantum Chemistry, Montreal, 1985.
2. R.F.W. Bader, *Atoms in Molecules: A Quantum Theory*, Oxford University Press, Oxford, U.K., 1990.
3. F.L. Hirshfeld, *Theor. Chim. Acta* **44**, 129 (1977).
4. C.A. Coulson, *Proc. Roy. Soc. (London)* **A169**, 413 (1939).
5. I. Mayer, *Simple Theorems, Proofs, and Derivations in Quantum Chemistry*, Kluwer Academic/Plenum Publishers, New York, 2003.
6. C.A. Coulson and G.S. Rushbrooke, *Proc. Roy. Soc. Edinburgh*, Sect. A **62**, 350 (1948).
7. T.G. Schmalz, W.A. Seitz, D.J. Klein, and G.E. Hite, *J. Am. Chem. Soc.* **110**, 1113 (1988).
8. V.F. Malichenko, *Moleculyarnie diagrammy organicheskych soedinenii*, Naukova Dumka, Kiev, 1982.
9. M.J.S. Dewar, *The Molecular Orbital Theory of Organic Chemistry.* McGraw-Hill, New York, 1969.
10. I. Mayer, Analytical Derivation of the Hückel "$4n+2$ rule." *Theor. Chem. Acc.*, **125**, 203–206 (2010).
11. R. Daudel, R. Lefebvre, and C. Moser, *Quantum Chemistry. Methods and Applications.* Interscience, New York, 1959.
12. F. Fratev, V. Enchev, P. Karadakov, and O. Castaño, *Int. J. Quantum Chem.* **26**, 993 (1984).
13. K. Baker, *Theor. Chim. Acta* **68**,221 (1985).
14. G. Lendvay, *J. Phys. Chem.* **98**, 6098 (1994).
15. G. Lendvay, T. Bérces, and F. Márta, *J. Phys. Chem. A* **101**, 1588 (1997).
16. G. Lendvay and B. Viskolcz, *J. Phys. Chem. A*, **102**, 10777 (1998).
17. P. Szabó, G. Lendvay, A. Horváth, and M. Kovács, *Phys. Chem. Chem. Phys.* **13**, 16033 (2011).
18. J.A. Pople and D.L. Beveridge, *Approximate Molecular Orbital Theory*, McGraw-Hill, New York, 1970.
19. K.R. Roby, *Mol. Phys.* **27**, 81 (1974).
20. I. Mayer and A. Hamza. *Int. J. Quantum Chem.* **103**, 798 (2005).
21. E. Clementi, *J. Chem. Phys.* **46**, 3842 (1967).
22. A. E. Clark and E. R. Davidson, *J. Chem. Phys.* **115** 7382 (2001).
23. A. E. Clark and E. R. Davidson, *Int. J. Quant. Chem.* **93**, 384 (2003).
24. I. Mayer, *Chem. Phys. Lett.* **97**, 270 (1983); addendum: Ref. 25.

25. I. Mayer, *Chem. Phys. Lett.* **117**, 396 (1985).
26. I. Mayer, *Int. J. Quant. Chem.* **23**, 341 (1983).
27. I. Mayer, The LCAO representation of the first order density matrix in non-orthogonal basis sets: a note. *J. Molec. Struct. (Theochem)* **255**, 1–7, (1992).
28. S.F. Vyboishchikov, P. Salvador, and M. Duran. *J. Chem. Phys.* **122**, 244110 (2005).
29. A. Hamza and I. Mayer, Physical analysis of the diatomic "chemical" energy components. *Theor. Chem. Accounts* **109**, 91–98 (2003).
30. I. Mayer and A. Hamza, *Int. J. Quant. Chem.* **92**, 147 (2003).
31. R. McWeeny, *Methods of Molecular Quantum Mechanics*, 2nd ed., Academic Press, London, 1992.
32. M.S. de Giambiagi, M. Giambiagi, and F.E. Jorge, *Theor. Chim. Acta* **68**, 337 (1985).
33. I. Mayer, *Chem. Phys. Lett.* **148**, 95 (1988).
34. P.-O. Löwdin, *J. Chem. Phys.* **18**, 365 (1950).
35. B.C. Carlson and J. M. Keller, *Phys. Rev.* **105**, 102 (1957).
36. I. Mayer, *Chem. Phys. Lett.* **393**, 209 (2004).
37. G. Bruhn, E.R. Davidson, I. Mayer, and A.E. Clark, *Int. J. Quant. Chem.* **106**, 2065 (2006).
38. E.R. Davidson, *J. Chem. Phys.* **46**, 3320 (1967).
39. I. Mayer, *Struct. Chem.* **27**, 51 (2016).
40. I. Mayer, *Chem. Phys. Lett.* **242**, 499 (1995).
41. I. Mayer, *Canadian J. Chem.* **74**, 939 (1996).
42. I. Mayer, *J. Phys. Chem.* **100**, 6249 (1996).
43. R. McWeeny, *Rev. Mod. Phys.* **32**, 335 (1960).
44. J.P. Foster and F. Weinhold, *J. Am. Chem. Soc.* **102**, 7211 (1980).
45. A.E. Reed, R.B. Weinstock, and F. Weinhold, *J. Chem. Phys.* **83**, 735 (1985).
46. A.E. Reed, L.A. Curtiss, and F. Weinhold, *Chem. Rev.* **88**, 899. (1988).
47. I. Mayer, Program "EFF-AO", Budapest 2008. May be downloaded from the web-site *http://occam.ttk.mta.hu* .
48. V. Magnasco and A. Perico, *J. Chem. Phys.* **80**, 971 (1967).
49. I. Mayer, G. Räther, and S. Suhai, *Chem. Phys. Lett.* **293**, 81 (1998).
50. R. Ponec and D.L. Cooper, *Faraday Discussions* **135**, 31 (2007).
51. I. Mayer, *Faraday Discussions* **135**, 128, 131 (2007).
52. D.L. Cooper and R. Ponec, *Phys. Chem. Chem. Phys.* **10**, 1319 (2008).
53. I. Mayer, Effective atomic orbitals: a tool for understanding electronic structure of molecules. *Int. J. Quantum Chem.* **114**, 1041–1047 (2014).
54. A. Schmiedekamp, D.W.J. Cruickshank, S. Skaarup, P. Pulay, I. Hargittai, and J.E. Boggs, *J. Am. Chem. Soc.* **101**, 2002 (1979).
55. N.C. Baird and K.F. Taylor, *J. Comp. Chem.* **2**, 225 (1981).
56. P.G. Mezey and E.-C. Haas, *J. Chem. Phys.* **77**, 870 (1982).

57. J. Janszky, R.H. Bartram, A.R. Rossi, and C. Corradi, *Chem. Phys. Lett.* **124**, 26 (1986).
58. K.B. Wiberg, *Tetrahedron,* **24**, 1083 (1967).
59. N.P. Borisova and S.G. Semenov, *Vestn. Leningrad. Univ.* **1973**, No. 16, 119.
60. K. Jug, Personal communication to I.M., 1985.
61. H. Fischer and H. Kollmar, *Theor. Chim. Acta* **16**, 163 (1970).
62. N.P. Borisova and S.G. Semenov, *Vestn. Leningrad. Univ.* **1976**, No. 16, 98.
63. I. Mayer, *Theor. Chim. Acta* **67**, 315 (1985).
64. I. Mayer, *Int. J. Quant. Chem.* **29**, 73 (1986).
65. M. Giambiagi, M.S. de Giambiagi D.R. Grempel, and C.D. Heyman, *J. Chim. Phys.* **72**, 15 (1975).
66. I. Mayer, Improved definition of bond orders for correlated wave functions. *Chem. Phys. Lett.* **544**, 83–86 (2012).
67. I. Mayer, Bond order and valence indices: a personal account. *J. Comput. Chem.* **28**, 204–221 (2007).
68. I. Mayer, DSc. Thesis, Budapest 1986.
69. W. Heitler and F. London, *Z. Physik* **44**, 455 (1927).
70. S. Weinbaum, *J. Chem. Phys.* **1**, 593 (1933).
71. D.R. Alcoba, R.C. Bochicchio, L. Lain, and A. Torre, *Phys. Chem. Chem. Phys.* **10**, 5144 (2008).
72. V.N. Staroverov and E.R. Davidson, *Chem. Phys. Lett.* **330**, 161 (2000).
73. I. Mayer, Energy partitioning schemes: a dilemma. *Faraday Discussions,* **135**, 439–450 (2007).
74. G. Schultz and I. Hargittai, *J. Phys. Chem.* **97**, 4966 (1993).
75. I. Mayer and Á. Gömöry, *J. Molec. Struct. (Theochem)* **311**, 331 (1994).
76. I. Mayer and Á. Gömöry, *Chem. Phys. Lett.* **344**, 553 (2001).
77. I. Mayer and Á. Gömöry, Program MNDO-MS, Budapest 1993/2001. May be downloaded from the web-site *http://occam.ttk.mta.hu* .
78. I. Mayer and A. Hamza, Program APOST-MS, Budapest, 2001. May be downloaded from the web-site *http://occam.ttk.mta.hu* .
79. Á. Somogyi, W.H. Wysocki, and I. Mayer, *J. Am. Soc. Mass Spectrom.* **5**, 704 (1994).
80. D.R. Armstrong, P.G. Perkins, and J.J.P. Stewart, *J. Chem. Soc., Dalton Trans.* **1973**, 838, 2273.
81. K. Takatsuka, T. Fueno, and K. Yamaguchi, *Theor. Chim. Acta* **48**, 175 (1978).
82. L. Lain, A. Torre, D.R. Alcoba, and R.C. Bochicchio, *Theor. Chem. Acc.* **128**, 405 (2011).
83. P. R. Surján, *Second Quantized Approach to Quantum Chemistry. An Elementary Introduction* Springer-Verlag, Berlin etc., 1989.
84. A. E. Clark and E. R. Davidson, *Mol. Phys.* **100**, 373 (2002).
85. A. E. Clark and E. R. Davidson, *J. Phys. Chem. A*, **106**, 6890 (2002).
86. E. R. Davidson and A. E. Clark, *Phys. Chem. Chem. Phys.* **8**, 1881 (2007).

87. D.R. Alcoba, A. Torre, L. Lain and R.C. Bochicchio, *J. Chem. Theory Comput.* **7**, 3560 (2011).
88. I. Mayer, Local spins: an alternative treatment for single determinant wave functions. *Chem. Phys. Lett.* **440**, 357–359 (2007).
89. I. Mayer, Local spins: improving the treatment for single determinant wave functions. *Chem. Phys. Lett.* **539–540**, 172–174 (2012).
90. P.-O. Löwdin *Phys. Rev.* **97** 1474 (1955).
91. P.-O. Löwdin, *Adv. Chem. Phys.* **2**, 207 (1959).
92. I. Mayer, Program BO-SPIN-2, Budapest, 2012. May be downloaded from the web-site *http://occam.ttk.mta.hu* .
93. I. Mayer, *Adv. Quantum Chem.* **12**, 189 (1980).
94. S. Angelov and I. Mayer, *Acta Phys. Hung.* **58**, 161 (1985).
95. E. Ramos-Cordoba, E. Matito, P. Salvador, and I. Mayer, Local spins: improved Hilbert-space analysis *Phys. Chem. Chem. Phys.* **14**, 15291–15298 (2012).
96. I. Mayer, *Izbrannye Glavy Kvantovoi Khimii. Dokazatel'stva teorem i Vivod Formul* Binom, Moscow 2006.
97. G. Tasi and I. Mayer, *Chem. Phys. Lett.* **449**, 221 (2007).
98. G. Tasi and D. Barna, *Int. J. Quant. Chem.* **109**, 2599 (2009).
99. E. Kapuy, C. Kozmutza, R. Daudel, and M.E. Stephens, *Theor. Chim. Acta* **53**, 147, (1079).
100. I. Mayer, Energy partitioning schemes. *Phys. Chem. Chem. Phys.* **8**, 4630–4646 (2006).
101. I. Mayer, Program BO-VIR, Budapest, 2005. May be downloaded from the web-site *http://occam.ttk.mta.hu* .
102. I. Mayer, On the promotion energy of an atom in a molecule. *Chem. Phys. Lett.* **498**, 366–369 (2010).
103. I. Mayer, *Struct. Chem.* **8**, 309 (1997).
104. R.J. Gillespie and I. Hargittai. *The VSEPR Model of Molecular Geometry*, Allyn and Bacon, Boston, 1991.
105. D. Asturiol, P. Salvador, and I. Mayer, *ChemPhysChem*, **10**, 1987 (2009).
106. N. Mott and I. Sneddon, *Wave Mechanics and Its Applications* Clarendon Press, Oxford 1948.
107. H.A. Bent, *Chem. Rev.* **61**, 275 (1961).
108. I. Mayer, *Int. J. Quant. Chem.* **70**, 41 (1998).
109. P. Salvador, D. Asturiol, and I. Mayer, *J. Comput. Chem.* **27**, 1505 (2006).
110. I. Mayer, G. Räther, and S. Suhai, *Chem. Phys. Lett.* **270**, 211 (1997).
111. I. Mayer, A chemical energy component analysis. *Chem. Phys. Lett.* **332**, 381–388 (2000).
112. I. Mayer and A. Hamza, *Theor. Chem. Acc.* **105**, 360 (2001).
113. I. Mayer and Á, Vibók, *Chem. Phys. Lett.* **140**, 558 (1987).
114. Z. Bikádi, G. Keresztury, S. Holly, O. Egyed, I, Mayer, and M. Simonyi, *J. Phys. Chem. A*, **105**, 3471 (2001).
115. I. Mayer, *Chem. Phys. Lett.* **382**, 264 (2003).

116. M.J.S. Dewar and W. Thiel, *J. Am. Chem. Soc.* **99** 4899 (1977).
117. I. Mayer, Improved chemical energy component analysis. *Phys. Chem. Chem. Phys.* **14**, 337–344 (2012).
118. I. Mayer, Program NEWENPART, Budapest, 2012. May be downloaded from the web-site *http://occam.ttk.mta.hu* .
119. R.K. Nesbet, *Adv. Chem. Phys.* **9**, 321 (1965).
120. P.Y. Ayala and G.E. Scuseria, *Chem. Phys. Lett.* **322**, 213 (2000).
121. P. Bultinck, C. Van Alsenoy, P.W. Ayers, and R. Carbó-Dorca, *J. Chem. Phys.* **126**, 144111 (2007).
122. I. Mayer and P. Salvador, Overlap populations, bond orders and valences for "fuzzy" atoms. *Chem. Phys. Lett.* **383**, 368–375 (2004).
123. A.D. Becke, *J. Chem. Phys.* **88**, 2547 (1988).
124. I. Mayer, unpublished results.
125. E. Matito, M. Solà, P. Salvador, and M. Duran, *Faraday Discussions* **135**, 325 (2007).
126. J.G. Ángyán, M. Loos, and I. Mayer, *J. Phys. Chem.* **98**, 5244 (1994).
127. I. Bakó, A. Stirling, A.P. Seitsosen, and I. Mayer, *Chem. Phys. Lett.* **563**, 97 (2013).
128. I. Mayer, I. Bakó, and A. Stirling, Are there atomic orbitals in a molecule? *J. Phys. Chem. A*, **115**, 12733–12737 (2011).
129. E. Ramos-Cordoba, P. Salvador, and I. Mayer, *J. Chem. Phys.* **138**, 214107 (2013).
130. I. Mayer and P. Salvador, *J. Chem. Phys.* **130**, 234106 (2009).
131. I. Mayer, *Chem. Phys. Lett.* **585**, 198 (2013).
132. P. Salvador, M. Duran, and I. Mayer, One- and two-center energy components in the AIM theory. *J. Chem. Phys.* **115**, 1153–1157 (2001).
133. *Chemical Concepts from Quantum Mechanics Faraday Discussions* vol. 135, RSC Publishing, 2007, pp 490-491.
134. P. Salvador and I. Mayer, Energy partitioning for "fuzzy" atoms. *J. Chem. Phys.* **120**, 5046–5052 (2004).
135. E. Ramos-Cordoba, E. Matito, I. Mayer, and P. Salvador, Toward a Unique Definition of the Local Spin. *J. Chem. Theory Comput.* **8**, 1270–1279 (2012).
136. H. C. Longuet-Higgins, p. 105 in *Quantum Theory of Atoms, Molecules and the Solid State* (ed. P.-O. Löwdin), Academic, New York, 1966.

Index